NATIONAL DEFENSE RESEARCH INSTITUTE

P9-BZH-984

First Steps Toward Improving DoD STEM Workforce Diversity

Response to the 2012 Department of Defense STEM Diversity Summit

Nelson Lim, Abigail Haddad, Dwayne M. Butler, Kate Giglio

Prepared for the Office of the Secretary of Defense

The research described in this report was prepared for the Office of the Secretary of Defense (OSD). The research was conducted within the RAND National Defense Research Institute, a federally funded research and development center sponsored by OSD, the Joint Staff, the Unified Combatant Commands, the Navy, the Marine Corps, the defense agencies, and the defense Intelligence Community under Contract W74V8H-06-C-0002.

Library of Congress Cataloging-in-Publication Data is available for this publication.

ISBN: 978-0-8330-8101-8

The RAND Corporation is a nonprofit institution that helps improve policy and decisionmaking through research and analysis. RAND's publications do not necessarily reflect the opinions of its research clients and sponsors.

Support RAND—make a tax-deductible charitable contribution at www.rand.org/giving/contribute.html

RAND® is a registered trademark

Cover image: iStockphoto/Thinkstock

© Copyright 2013 RAND Corporation

RAND OFFICES
SANTA MONICA, CA • WASHINGTON, DC
PITTSBURGH, PA • NEW ORLEANS, LA • JACKSON, MS • BOSTON, MA
DOHA, QA • CAMBRIDGE, UK • BRUSSELS, BE
www.rand.org

Preface

On August 18, 2011, the President issued Executive Order 13583, "Establishing a Coordinated Government-Wide Initiative to Promote Diversity and Inclusion in the Federal Workforce." The Department of Defense (DoD), in response, soon released an organizational diversity and inclusion strategic plan addressing workforce diversity, workplace inclusion, and force sustainability. In turn, leaders from DoD's Assistant Secretary of Defense for Research and Engineering (ASD[R&E]) and the Office of Diversity Management and Equal Opportunity (ODMEO) sought to further support the President's order in ways specific to DoD's science, technology, engineering, and mathematics (STEM) workforce, and convened a two-day conference, the Summit, on November 1, 2012. Senior executives working on personnel issues within DoD, federal agencies, the private sector, and academia gave presentations and participated in discussions. These presentations focused on issues related to the DoD STEM workforce and its diversity, particularly racial/ethnic and gender diversity.

This report supports the efforts of the DoD Diversity STEM Summit by responding to the proceedings in a way that may provide a foundation for further research, analysis, and action. The report describes current diversity policies and demographic trends and provides a template for comparing STEM-diversity outreach programs to support ASD(R&E) and ODMEO's interest in bringing populations currently underrepresented in STEM into DoD's STEM workforce mix. The report also offers a number of initial recommendations for DoD leaders to consider as they move forward with their efforts to increase the diversity of the STEM workforce.

This research was sponsored by ASD(R&E) and ODMEO and conducted within the Forces and Resources Policy Center of RAND's National Defense Research Institute, a federally funded research and development center sponsored by the Office of the Secretary of Defense, the Joint Staff, the Unified Combatant Commands, the Department of the Navy, the Marine Corps, the defense agencies, and the defense Intelligence Community. For more information on the RAND Forces and Resources Policy Center, see http://www.rand.org/nsrd/ndri/centers/frp.html or contact the director (contact information is provided on the web page).

Contents

Figures

Tables

Summary

On August 18, 2011, the President issued Executive Order 13583, "Establishing a Coordinated Government-Wide Initiative to Promote Diversity and Inclusion in the Federal Workforce." The Department of Defense (DoD), in response to the order, soon released an organizational diversity and inclusion strategic plan addressing workforce diversity, workplace inclusion, and force sustainability. In turn, leaders from DoD's Research and Engineering (ASD[R&E]) and the Office of Diversity Management and Equal Opportunity (ODMEO) sought to further support the President's order in ways specific to DoD's science, technology, engineering, and mathematics (STEM) workforce. These leaders convened a two-day conference, the DoD Diversity STEM Summit, on November 1, 2012.

Over the course of the Summit, senior executives from DoD, federal agencies, the private sector, and academia presented on and discussed issues related to the diversity of the DoD STEM workforce. Participants described existing initiatives and proposed new ideas to increase STEM participation overall and of underrepresented groups specifically.

ODMEO asked researchers from the RAND Corporation to support Summit efforts by responding to the proceedings with actionable recommendations. Over the course of the Summit, several key questions emerged:

- How can DoD begin to better position itself to establish a diverse STEM workforce?

- What do demographic trends suggest about DoD's current STEM workforce?
- What are DoD and partners doing to increase diversity in the STEM workforce and what else can be done?

The RAND team synthesized the information presented at the Summit as well as conversations following each presentation to answer these questions and offer a number of possible process-related recommendations that DoD can take as a first step toward its STEM workforce diversity goals.

How Can DoD Begin to Better Position Itself to Establish a Diverse STEM Workforce?

A number of STEM- and diversity-related policies have gone into effect over the past decade. For example, the 2007 America Creating Opportunities to Meaningfully Promote Excellence in Technology, Education, and Science Act, or America COMPETES Act (P.L. 110-69), broadly targets the nation's current and projected need for more STEM workers to effectively respond to a globally competitive world. America COMPETES was reauthorized in 2010 (P.L. 111-358) to provide funding for research, development, and education in STEM areas until 2013. Additionally, the intention to increase minority representation across all federal organizations, including DoD, is articulated in Executive Order 13583. The order instructs leaders of federal agencies to make changes in the ways in which minority groups are recruited, hired, promoted, and retained. STEM skills are specifically referred to in the U.S. Office of Personnel Management, Office of Diversity and Inclusion's *Government-Wide Diversity and Inclusion Strategic Plan 2011* and in the *DoD STEM Strategic Plan: FY 2013–FY 2017*.

Our initial review of these policies suggests they are being driven in part by several assumptions:

- That there is a need for talented and innovative STEM workers to meet 21st century global challenges.

- That the nation's demographic makeup is changing and will continue to change.
- That the federal workforce, including DoD, must be inclusive and reflect the demographics of the country it serves.

This report does not evaluate the first and third assumptions, but the review of current STEM- and diversity-related national and DoD policy suggests that there are substantial grounds on which to base further action. ASD(R&E) and ODMEO may consider leveraging these policies separately or together as the organizations move forward in developing the diversity of the STEM workforce.

What Do Demographic Trends Suggest About DoD's Current STEM Workforce?

The Census Bureau's 2008 National Population Projections demonstrate that the United States is in the midst of a major demographic shift. These projections show that the nation's population is becoming significantly more Hispanic and less white, non-Hispanic. Figure S.1 graphically illustrates this shift: The Census Bureau estimates that in 2000, whites not of Hispanic origin (referred to hereafter as whites) made up about 70 percent of 18–65-year-olds, and the bureau projects that in 2050 they will make up less than 45 percent, with Hispanics increasing from 12 percent to 30 percent. The proportion of Asians is also projected to increase, from about 4 percent in 2000 to about 8 in 2050.[1] These projections rely on various assumptions about future mortality, birth rates, and migration.

Figure S.2 shows the current composition of 23–29-year olds, the proportion of those who have college degrees in a STEM field, and the proportion of those who have college degrees (in any field) and work in a STEM occupation. Notably, Hispanics make up 20 percent of the

---

[1] The "other" group, in the census data as well as in the American Community Survey (ACS) data shown later in the report, includes both non-Hispanics who are multiracial and non-Hispanics who are American Indian, Alaskan Natives, or Native Hawaiian or other Pacific Islander.

Figure S.1
Census Projections, 2000–2050, 18–65-Year-Olds

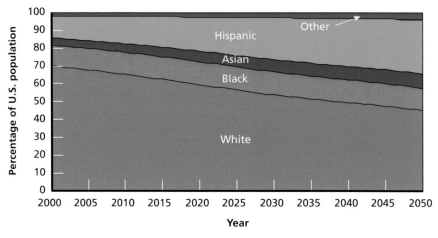

SOURCE: Based on authors' computations with 2008 Census projections data (U.S. Census Bureau, 2008).

RAND RR329-S.1

Figure S.2
2010 Percentage of 23–29-Year-Olds in Overall Population, Among STEM Degree Holders, and Among Those with College Degrees in STEM Occupations, by Race/Ethnicity

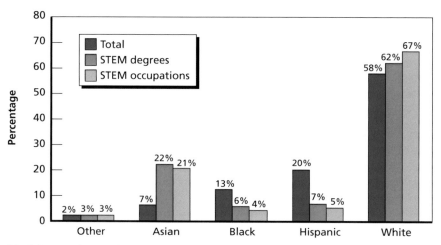

SOURCE: Based on authors' computations with 2010 ACS data (Ruggles et al., 2010).

RAND RR329-S.2

overall young adult population but only 7 percent of those with degrees in STEM fields, and only 5 percent of those with a college degree who work in a STEM occupation. Similarly, black young adults make up 13 percent of the overall population but only 6 percent of those with degrees in STEM fields, and only 4 percent of those with a college degree who work in a STEM occupation.

The significantly lower representation of Hispanics among those with a STEM degree and among those in the STEM workforce, coupled with the rapidly growing Hispanic population, presents a major challenge to DoD's efforts to make its STEM workforce racially representative of the nation. Without significant convergence among rates of STEM participation by racial/ethnic group, the gap between the proportion of Hispanics in the working-age population and the proportion of STEM workers who are Hispanic will grow significantly.

Currently, the DoD STEM workforce closely parallels the citizen STEM workforce in terms of racial/ethnic composition, indicating that the factors affecting the composition of the overall STEM workforce are also affecting DoD STEM hiring (see Figure S.3).[2] This similarity suggests that it may be difficult for DoD to hire a STEM workforce that is significantly more racially diverse than the overall STEM workforce.

What Are DoD and Partners Doing to Increase Diversity in the STEM Workforce, and What Else Can Be Done?

Our initial comparison of DoD STEM outreach activities conducted across the United States suggests that the goals and intended participants of these efforts vary greatly. Some of the programs aim to get STEM students into the DoD hiring pipeline, while others have

---

[2]   The overall STEM workforce, using the DoD definition, which includes health practitioners, is more than half female, whereas the DoD civilian STEM workforce is only 29 percent female. We attribute this to the fact that health practitioners make up over half of citizen STEM workers, and women make up a large majority of that group. In contrast, we believe that health practitioners constitute much less than half of the DoD STEM workforce, although we do not have data to support this speculation.

Figure S.3
Overall STEM Citizen Workforce and DoD Civilian STEM Workforce, by Race/Ethnicity

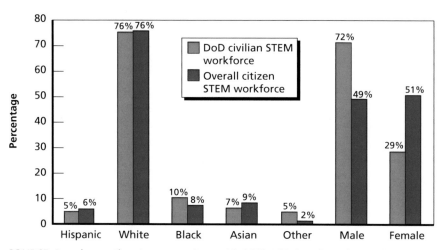

SOURCE: Based on authors' computations with 2010 ACS data (Ruggles et al., 2010) and with DoD-provided data.
NOTE: For an explanation of the methodology behind this graph, including the gender disparities between the DoD STEM workforce and the overall STEM workforce, see Chapter Three.
RAND RR329-S.3

broader goals, such as community outreach. Many efforts provide educational support. Programs such as SeaPerch, the Navy's robotics competition, are designed to increase school-age children's interest in STEM and problem-solving skills. Other educational programs, such as the Air Force's Teachers Materials Camp, supports school curriculum development. Several programs directly "feed" into the DoD workforce. Notable in this category is the Science, Mathematics, And Research for Transformation (SMART) Scholarship program, which awards scholarships to students in certain STEM fields in return for postgraduate employment with DoD.

However, review of the programs suggests a need for greater goal articulation and alignment as well as program assessment. Program goals, associated costs, and minority participation were unclear or unknown to presenters and to the program leaders we consulted with after the summit, or this information was just not shared with us. From the information we did gather, we found that (1) program goals may overlap in some cases and (2) there is no uniform measurement used to determine how successful these programs are in reaching out to different demographic groups or in achieving other goals, such as improved academic performance or pipelines to federal service. Once the DoD STEM diversity goals are articulated, assessment will be necessary to determine which programs are achieving desired outcomes.

Recommendations

Given that this report is only an initial response to Summit proceedings, we recommend first and foremost that DoD clearly articulate which aspects of diversity it wishes to prioritize and establish a common set of specific goals to pave the way toward reaching desired outcomes. After these outcomes are expressed, the organization will be better positioned to analyze whether current practices are effective and to explore changes to make practices more effective. Over the course of the Summit, for example, speakers focused primarily on racial/ethnic diversity, and gender diversity to a lesser extent. In response, in this report we focus primarily on the representation of racial/ethnic and gender categories in the STEM workforce and how current DoD outreach programs target those groups. However, other categories of diversity, such as religious background, national origin, and skill sets, are included in definitions of diversity throughout various DoD policy documents. These categories were seldom mentioned over the course of presentations and discussions, which is reflected in our response. Briefly stated, a clearer articulation of diversity goals is necessary to maximize the effectiveness of later steps.

The initial findings, especially those related to program assessment, suggest a second overarching recommendation: that DoD work toward coordinating efforts across the organization to reach its STEM-diversity workforce goals. By *coordination*, we refer to the synchronization of organizational efforts, including the efforts of DoD as well as supporting agencies and external stakeholders, to improve effectiveness and efficiency and reduce costs through program sharing and minimization of overlap.

We offer the following managed-change plan to bring together the various efforts designed to promote a diverse workforce, including the STEM workforce. We recommend that DoD take action incrementally and in order: in the short term (1–12 months), mid term (1–3 years), and long term (4+ years).

Short Term
Recommendation 1: Articulate DoD STEM and diversity goals and align policies and practices within the DoD STEM community toward achievement of these goals.
DoD's current definition of diversity is broad and, as such, is useful for defining diversity according to organizational strategy:

> Diversity is all the different characteristics and attributes of the DoD's Total Force, which are consistent with our core values, integral to overall readiness and mission accomplishment, and reflective of the nation we serve. (U.S. Department of Defense, Science, Technology, Engineering, and Mathematics [STEM] Executive Board, 2012, p. 3)

However, the definition needs to be translated into a more specific tactical definition that DoD STEM leaders can use for designing policy and setting goals. It may not be possible to effectively focus on every possible aspect of diversity implied or explicitly mentioned in DoD policy documents. Relatedly, the Office of the Secretary of Defense and the services have many programs without clearly articulated goals or assessment of their returns on investment. In the new era of fiscal responsibility, it is essential that members of the DoD STEM community come together under a set of clear strategic objectives, including operationalizing what they mean by diversity, articulating clear goals, and focusing their resources on meeting those goals. In this report, we offer an option for collecting information to compare various efforts, focusing largely on racial/ethnic diversity. This comparison of different efforts in terms of goals, intended participants, and other factors may help improve overall efforts as well as reduce overlap.

Recommendation 2: Establish closer working relationships between ASD(R&E) and ODMEO in the short term and between AT&L and P&R in the long term.

Both ASD(R&E) and ODMEO recently finalized and published DoD strategic plans in their policy domains (STEM and diversity, respectively). Currently, both organizations are developing their implementation plans, which is the opportune time for both organizations to closely align their plans to maximize the impact of their efforts. ASD(R&E) and ODMEO should consider building partnerships between the larger DoD organizational components, Acquisition, Technology, and Logistics (AT&L) and Personnel and Readiness (P&R), for maximum impact.

Recommendation 3: Focus on building a pipeline to DoD employment.

Both ASD(R&E) and ODMEO sponsor a number of outreach activities and internships. They and their service partners should consider co-sponsoring outreach activities and coordinating internships for better return on investment. More importantly, DoD should develop closer links from outreach to internships, from scholarship to hiring.

Mid Term

Recommendation 4: Expand strategic initiatives to include the Total Force.

As DoD implements its STEM strategic plan, it should consider expanding the scope of certain programs to include all members of the Total Force—including not only active and reserve components, but also civil servants and contractors. Identifying programs from one component that could be successfully implemented in other components may help DoD to realize its STEM strategic goal.

Recommendation 5: Engage the Military Personnel Policy (MPP) and Civilian Personnel Policy (CPP) offices to overhaul recruiting of STEM professionals for all components.

The National Academies expressed concerns in 2012 that DoD is not the employer of choice among the most talented STEM professionals (National Research Council, 2012, pp. 116–117). AT&L and

P&R should consider whether transforming aspects of DoD personnel policies could help DoD to achieve its STEM goals. ASD(R&E) and ODMEO may serve as catalysts for this organizational transformation, which will prepare DoD to meet emerging STEM needs and become an employer of choice of top STEM talents, and also engage with other DoD offices that set and implement personnel policy.

Recommendation 6: Establish specific goals for the representation of minorities and women in the STEM applicant pool.

Having goals is an important part of organizational change efforts. To meet its goal of increasing the representation of minorities and women among STEM workers, we recommend that DoD articulate specific outcomes so that leaders will know what to work toward and DoD will be able to self-evaluate and change its strategy and goals in response to results. Because of legal limitations, setting goals for the number of hires or employees is not feasible. However, one potential option is to set goals for the applicant pool for DoD jobs and programs. For example, the goal might be that the applicant pool for specific jobs mirror the demographics of its occupation in the overall labor market, perhaps based on the sort of census data analysis we perform in this report. It is possible that those standards are already being met in terms of racial/ethnic diversity, since the racial/ethnic composition of STEM workers within DoD already mirrors fairly closely the racial/ethnic composition of citizen STEM workers in the overall labor market. Another potential goal might be that the applicant pool for specific jobs mirror some weighted average of the demographics of the overall working-age population and the relevant labor market. We are not recommending a particular goal, but rather suggesting that DoD develop some type of measurable outcome, which will allow internal or external stakeholders to evaluate progress toward the articulated goals.

Long Term
Recommendation 7: Establish formal ties between policies and practices of AT&L and P&R (ASD[R&E] and ODMEO).

Improving the diversity of the DoD STEM workforce cannot be done overnight. To see measurable changes, DoD must institutionalize closer coordination and collaboration between ASD(R&E) and ODMEO, as well as create innovative policies and practices that streamline recruiting, hiring policies, and practices between AT&L and P&R.

Going Beyond DoD

Short Term

Recommendation 8: Consider establishing the Defense Diversity and STEM Advisory Council, representing the defense STEM "ecosystem," with expanded mandate to provide oversight and advise the Secretary of Defense.

In the course of the Summit, presenters and audience members referred to a STEM "ecosystem" that includes DoD and "the entire system supporting it, including the labs and industry." It was suggested this ecosystem could produce opportunity and enhance diversity and inclusion, and as such yield "hybrid forms of new innovation" to ensure the nation's competitiveness. We suggest that DoD consider establishing a Defense Diversity and STEM Advisory Council that has expertise in a number of personnel-related areas, such as recruitment, diversity management, and STEM training. Such a council can give recommendations and feedback as the organization works toward fulfilling its workforce goal.

Recommendation 9: Be an agent for a national campaign.

DoD should consider acting as a catalyst to reach its own goal by engaging with external stakeholders in a national campaign to improve the diversity of its STEM workforce. Given DoD's size and visibility, as well as the centrality of STEM to its overall mission and its significant interactions with other STEM employers within both the government and the private sector, the organization may be able to influence a national campaign. Before engaging in a full campaign, however, analysis on the cost-effectiveness of support and participation should be conducted.

Mid Term

Recommendation 10: Work with industry and academia to increase diversity within STEM professions.

The National Academies (2012) pointed out the importance of the industrial suppliers of DoD to meet the STEM challenges. Defense industry leaders such as Lockheed Martin presented evidence of success at the Summit. We suggest that DoD consider working together with industry and academia to increase diversity within STEM professions overall in order to fulfill its workforce goal.

Recommendation 11: Support and track success of national campaign to improve diversity within the STEM workforce.

In the medium term, DoD should continue taking a role in promoting STEM careers and education with other partners to increase the diversity of the national STEM workforce. Measuring and tracking the success of STEM-diversity programs will also be important for evaluating the performance of those programs and revamping them as needed.

Long Term

Recommendation 12: Enable the Defense Diversity and STEM Advisory Council to monitor policies and practices to increase the diversity of the STEM workforce.

The effort to improve the diversity of DoD's STEM workforce must endure over the long term, and thus the Defense Diversity and STEM Advisory Council should be empowered to monitor and assess DoD efforts and advise the Secretary of Defense on how to improve STEM policies and practices.

Recommendation 13: Sustain efforts to improve the diversity of the overall STEM workforce.

In the long run, DoD may be able to meet its STEM-diversity goals by sustaining efforts to increase diversity within its own STEM workforce, as well as continuing to contribute to national efforts to improve the diversity of the overall STEM workforce.

Acknowledgments

The authors wish to thank Reginald Brothers and Clarence Johnson for giving us an opportunity to support their effort in aligning DoD STEM and diversity strategic goals. We thank Karen Harper for making the vision of the Summit a reality. The collaborative effort between AT&L and P&R can significantly improve the efficiency and effectiveness of STEM and diversity initiatives. We also thank all participants of the 2012 DoD STEM Diversity Summit, especially Laura Adolfie, director, STEM Development Office, and Laura Stubbs, director, S&T Initiatives, for providing critical information about DoD's STEM initiatives. The information they provided is the backbone of this report. Finally, we thank RAND colleagues Kirsten Keller and Beth Asch for their insightful comments on an earlier draft of this report. Catherine Chao provided administrative support to the RAND research team, and we offer our thanks to her as well.

Abbreviations

ACS	American Community Survey
ASD(R&E)	Assistant Secretary of Defense for Research and Engineering
AT&L	Acquisition, Technology, and Logistics
BLS	Bureau of Labor Statistics
CPP	Civilian Personnel Policy
DoD	U.S. Department of Defense
DoDEA	Department of Defense Education Activity
HBCUs	Historically Black Colleges and Universities
MLDC	Military Leadership Diversity Commission
MPP	Military Personnel Policy
MSI	minority-serving institution
ODMEO	Office of Diversity Management and Equal Opportunity
OSD	Office of the Secretary of Defense
P&R	Personnel and Readiness
S&E	science and engineering
SMART	Science, Mathematics & Research for Transformation

SOC   Standard Occupational Classification

STEM   science, technology, engineering, and mathematics

CHAPTER ONE

Introduction

> One third of DoD jobs are STEM-related. There's a demographic change. We're in a battle for talent. We need to get the best and brightest our nation has to offer. (Clarence Johnson, DoD STEM Diversity Summit, November 2012)

On August 18, 2011, the President issued Executive Order 13583, "Establishing a Coordinated Government-Wide Initiative to Promote Diversity and Inclusion in the Federal Workforce." The order focuses on changing how the government recruits, hires, trains, promotes, and retains its workforce in order to benefit from the talents of all parts of society. The Department of Defense (DoD) Research and Engineering Enterprise and the DoD Office of Diversity Management and Equal Opportunity (ODMEO) have taken steps toward enacting the order by developing plans to increase the diversity of DoD's science, technology, engineering, and mathematics (STEM) workforce. However, creating and sustaining institutional change is a complex process, and, to generate ideas, the Assistant Secretary of Defense for Research and Engineering (ASD[R&E]) and ODMEO convened a two-day conference, the DoD Diversity STEM Summit. On November 1 and 2, 2012, senior executives and experts from across DoD, the services, federal agencies, the private sector, and academia explored what changes might be needed to effectively and efficiently achieve diversity goals across the DoD STEM workforce. Presenters and panelists included senior executives working on personnel issues within DoD and federal agencies, as well as senior leaders working on similar issues from the private sector and academia.

A Time for Addressing Challenges and Finding Solutions

A number of Summit presentations and discussions described and proposed initiatives to increase STEM participation overall and particularly among demographic groups who are underrepresented in the STEM workforce and DoD STEM workforce, relative to the labor force as a whole. These groups include women, blacks, and Hispanics. Other panel discussions examined efforts and related challenges that DoD faces in attracting and retaining "the best and brightest" STEM talent from all demographic groups. The Summit agenda is presented in Appendix A.

Reginald Brothers, Deputy Assistant Secretary of Defense for Research, opened the Summit by reminding participants of its goals: increasing diversity throughout DoD and particularly in the STEM fields. He emphasized that diversity includes more than demographic differences by stating, "When we think about diversity, we need to think of it across people, but also across disciplines." According to Brothers, the ultimate goal of improving the diversity of DoD's STEM workforce is to find innovative solutions for evolving national security challenges. Many of the solutions, remarked Brothers, may need to cut across traditional disciplinary boundaries. Brothers reminded the panelists and audience members that the primary objective of the Summit is to identify actionable recommendations to move forward.

Clarence Johnson, principal director of ODMEO, also provided opening remarks to set the agenda for discussion. Reinforcing Brothers's comments, he emphasized that DoD currently defines diversity as

> all the different characteristics and attributes of the DoD's Total Force, which are consistent with our core values, integral to overall readiness and mission accomplishment, and reflective of the nation we serve.

Johnson suggested to the audience that "diversity is a strategic imperative, critical to mission readiness and accomplishment and a leadership requirement." He urged participants of the Summit to find ways to ensure leadership commitment to an accountable and sustained diversity effort, employ an aligned strategic outreach effort to identify, attract, and recruit from a broad talent pool reflective of the best of the nation that we serve, and develop, mentor, and retain top talent from across the Total Force.

Purpose of This Report

Researchers from the RAND Corporation were asked to support the Summit and ASD(R&E) and ODMEO by responding to the proceedings with recommendations for further action. RAND analysts attended the two-day event and engaged in discussion with participants.

Over the course of the Summit, several key questions emerged:

- How can DoD begin to better position itself to establish a diverse STEM workforce?
- What do demographic trends suggest about DoD's current STEM workforce?
- What are DoD and partners doing to increase diversity in the STEM workforce and what else can be done?

In this report, we synthesize the information presented at the Summit as well as conversations following each presentation to answer these questions and offer a number of possible process-related recommendations for DoD to consider as first steps toward articulating and meeting their STEM workforce diversity goals.

Our conference notes can be reviewed in Appendix B. In addition, we reviewed DoD STEM and diversity strategic plans, national and DoD workforce statistics, and relevant background literature. This literature includes the recent report by the National Academies, *Assuring the U.S. Department of Defense a Strong Science, Technology, Engineering, and Mathematics (STEM) Workforce* (2012), which was cited by several speakers over the course of the Summit.

Limitations of This Report

It is important to address the exploratory nature of this report, which reflects the goals of the Summit. Our response was needed one month after the Summit, giving the team a short time to review materials additional to the meeting notes; our sources include current legislation, DoD's current STEM strategic plan, demographic trends, and some responses to additional queries we sent to presenters. This report suggests that these materials are the grounds to begin a coordinated effort in building a U.S.-based diverse DoD STEM workforce. However, the effort in reaching this goal may require deeper research and analysis in the areas of organizational management, STEM education, program and educational assessment and metrics, and more. Additionally, we make no arguments regarding the value of a diverse STEM workforce: The goal of this report is to assist DoD with thinking about how to meet its goal of a diverse STEM workforce, not to evaluate that goal. Overall, while the study may be useful in providing grounds for change, it is clear that more detailed and goal-oriented research should be undertaken to fulfill ASD(R&E) and ODMEO's goal of building and sustaining a diverse STEM workforce.

Organization of This Report

The report is organized according to the three key questions presented above. In Chapter Two, we review current national priorities and DoD strategic plans to find a potential strategic alignment between ASD(R&E) and ODMEO. In Chapter Three, we review the demographic trends that may affect the diversity of the STEM workforce. In Chapter Four, we briefly review STEM-related programs and initiatives described over the course of the Summit. In Chapter Five, we present options and recommendations for DoD leaders and partners to consider as they move forward with their STEM-diversity goals.

Three appendixes support this report. Appendix A presents a copy of the Summit agenda, while Appendix B offers a copy of the Summit presentation and conversation notes. Appendix C, provided on the web at http://www.rand.org/pubs/research_reports/RR329.html, provides copies of Summit presentations by representatives of the Navy, Air Force, Army, Reserve Affairs, ODMEO, and the Office of the Under Secretary of Defense for Personnel and Readiness.

CHAPTER TWO

Review of Policies and Strategies Aiming to Align DoD STEM and Diversity Goals with National Priorities

For the United States to maintain the global leadership and competitiveness in science and technology that are critical to achieving national goals today, we must invest in research, encourage innovation, and grow a strong, talented, and innovative science and technology workforce. (National Academies, 2007, p. 1)

A national effort to sustain and strengthen S&E [science and engineering] must also include a strategy for ensuring that we draw on the minds and talents of all Americans, including minorities who are underrepresented in S&E and currently embody a vastly underused resource and a lost opportunity for meeting our nation's technology needs. (National Academies, 2011, p. 2)

A STEM-capable workforce and workforce diversity are high-priority policy areas for DoD. There have been a number of initiatives, programs, and policies regarding both in the past decade, but it is in the past three years especially that STEM and diversity have been addressed in tandem.

In this chapter, we briefly review two recent policies related to STEM and diversity, respectively. We also review the DoD diversity strategic plan, created in response to Executive Order 13583 and a recent version of DoD's STEM strategic plan. The purpose of this review is to show the ways in which diversity and STEM have been concurrently addressed in recent years. One of our conclusions from this review is that, although diversity has been articulated as a goal by several policy documents, the definition of diversity has been so general that it is not possible to evaluate how well the DoD is currently performing. Therefore, it is necessary for DoD to articulate a more tactical definition that military leaders can reference when creating programs and evaluating performance.

STEM-Related Policies Begin to Target Workforce Diversity

A high-quality STEM workforce has been formally recognized as necessary to national security since at least 1947, when the Research and Development Board formed within DoD (U.S. Department of Defense, Research and Engineering Enterprise, 2012). More recently, both the executive and legislative branches of government have demonstrated the priority placed on the STEM workforce by a series of legislations: the American Competitive Incentive Act, the America COMPETES Act, and substantial appropriations through the American Recovery and Reinvestment Act of 2009 (National Academies, 2011). Here we review the America COMPETES Act, as it is targeted toward the need to further develop national STEM capability and calls upon the scientific community to address STEM workforce diversity. We also review DoD's current STEM strategic goals, as diversity is addressed in its workforce-development objectives.

America COMPETES Act Targets Diversity Through Research Directives

The America Creating Opportunities to Meaningfully Promote Excellence in Technology, Education, and Science Act, or America COMPETES Act (P.L. 110-69), broadly targets the nation's current and projected need for more STEM workers to effectively respond to a globally competitive world. The act was established in part in response to a 2007 National Academies report, *Rising Above the Gathering Storm: Energizing and Employing America for a Brighter Economic Future*, which argued that the state of STEM fields was at a critical junction; investment in an innovative STEM agenda and a strong, talented workforce must be made, the report argued, lest the nation fall behind in STEM areas in a competitive global market. The America COMPETES Act thus directs a range of government science-related agencies and offices to cooperate with one another and to invest in high-risk, high-reward research areas and industrial opportunities, and the act promotes educational reforms in order to "invest in innovation through research and development, and to improve the competitiveness of the United States" (P.L. 110-69). The Act was reauthorized in 2010 (P.L. 111-358) and provides funding for these areas until 2013.

As part of the act, the National Academy of Sciences was charged to investigate barriers to increasing the number of underrepresented minorities in STEM fields and to "identify strategies for bringing more underrepresented minorities" into the STEM workforce (P.L. 110-69 § 5003). In response, the National Academies released the 2011 report, *Expanding Underrepresented Minority Participation: America's Science and Technology Talent at the Crossroads*. Experts from the National Academy of Sciences, the National Academy of Engineering, and the Institute of Medicine argued:

> Critical issues for the nation's S&E infrastructure remain unsettled. Among them, America faces a demographic challenge with regard to its S&E workforce: Minorities are seriously underrepresented in science and engineering, yet they are also the most rapidly growing segment of the population. (National Academies, 2011, p. 1)

For many years, the U.S. STEM workforce has been "predominately male and overwhelmingly white and Asian," according to the report, while "[n]on-U.S. citizens, particularly those from China and India, account[ed] for almost all growth in STEM doctorate awards" (p. 22). The study presents several reasons why reliance upon the workforce pipeline, as currently established, may no longer be feasible:

- **The U.S. white population is increasing at a slower rate than the total U.S. population.** The 2011 National Academies report cites the 2010 U.S. Census results, suggesting that "while the white alone population increased numerically over the [past] 10-year period, its proportion of the total population declined from 75 percent to 72 percent" (p. 4). The U.S. population is becoming a more-varied mix that includes fast-growing Asian and Hispanic populations.

- **Immigration laws and interest by noncitizens in working in the United States are changing.** According to the 2011 report, non–U.S. citizens are expressing less interest in earning graduate degrees in the United States as other nations experience higher education reforms and as U.S. student visa laws have become more stringent after the events of September 11, 2011. Further, those that do earn STEM degrees in the United States do not necessarily remain in the United States to work. Additional obstacles come with reliance upon a non-U.S. STEM workforce. According to the 2012 National Academies report focused specifically on the DoD STEM workforce, constraints on the H1-B high-tech visa and long wait times for permanent workforce "green cards" are already posing problems for many. Also, much DoD-related STEM work, especially in the National Laboratories, requires security clearances that are only issued to U.S. citizens.

In the 2012 National Academies report, the authors argue that for STEM-related national security goals to be reached in the future, the United States must develop a STEM workforce that reflects U.S. population trends.

There is no immediate national STEM or DoD STEM workforce shortage in sight, except in newer areas such as cyber security. Yet there is growing demand for STEM talent in a variety of fields, including from DoD contractors (National Academies, 2012). Additionally, the Congressional Budget Office has found that federal workers with professional or doctoral degrees earn about 18 percent less (wages and benefits) than those in the private sector (Falk, 2012). This difference reflects a general tendency for the federal government to have a compressed pay scale, where workers with less human capital tend to earn more than they would in the private sector and workers with significant amounts of human capital tend to earn less. As a result, DoD may have difficulty recruiting the "the best and brightest" now and in the coming years, and as emerging fields such as cyber security demonstrate, it is difficult to predict when and in what areas STEM talent will be needed.

DoD's Current STEM Strategic Plan Targets Diversity Broadly

DoD's *STEM Strategic Plan, FY 2013–FY 2017* was released in draft form on September 5, 2012. The goal of the plan is to "ensure that the Department has the STEM expertise necessary to develop technological solutions in an ever-changing threat environment and affords DoD Components the ability to tailor their approach to achieve these objectives" (p. 1). Developing an appropriate workforce is the first goal of three offered in the document; the opening statement emphasizes a need to attract and develop "a diverse, world-class STEM talent pool and workforce with the creativity and agility to meet national defense needs" (p. 1). In the 2012 *DoD Diversity and Inclusion Strategic Plan*, diversity is defined as "all the different characteristics and attributes of the DoD's Total Force, which are consistent with our core values, integral to overall readiness and mission accomplishment, and reflective of the nation we serve" (p. 3). This definition is too general to be useful in terms of designing or evaluating policy: A more specific focus is necessary.

However, this broad definition of diversity corresponds with the 2011 findings of the congressionally mandated Military Leadership Diversity Commission (MLDC). While the work of the MLDC focuses specifically on leadership of the Armed Forces and not the Total Force, their definition draws on categories from workforce management literature:

- **Demographic diversity** includes immutable differences among individuals, such as race, ethnicity, gender, and age, as well as differences in personal background, such as religion, education level, and marital status. This is the traditional definition of diversity.
- **Cognitive diversity** includes different skill sets; personality types, such as extrovert/introvert; and different thinking styles,

such as quick and decisive versus slow and methodical (see Riche, Kraus, and Hodari, 2007).

- **Structural diversity** for DoD includes civilian and military members' organizational background, such as their service, department, component (active, reserve, or civilian), and work function.
- **Global diversity** includes affiliations with nations other than the United States.

As with the diversity definition in the DoD strategic plan, this definition is extremely broad. In order to set goals and assess whether they are being met, DoD must focus on particular categories of diversity for their STEM workforce. This report focuses on demographic diversity, particularly race/ethnicity and gender, because these issues were the focus during the Summit and, to the best of our knowledge, have been the focus of most of DoD's diversity efforts.

DoD Diversity Policy Focuses on STEM for Mission Effectiveness

In the introduction to this report, we referenced Executive Order 13583, "Establishing a Coordinated Government-Wide Initiative to Promote Diversity and Inclusion in the Federal Workforce." The Summit, as noted, was designed to respond to this directive to increase diversity throughout all agencies and components of the government. On August 18, 2011, the President issued the order as a way to bring together and emphasize a number of earlier diversity-related executive orders, including Executive Order 13171, "Hispanic Employment in the Federal Government" (October 2000), and Executive Order 13548, "Increasing Federal Employment of Individuals with Disabilities" (July 2010). Executive Order 13583 affirms the importance of attaining a diverse and qualified federal workforce and instructs leaders of federal agencies to improve practices for promoting diversity and inclusiveness in their workforce. No definition of diversity is given.

STEM skills are not addressed specifically by the order, but they are referred to in the follow-on document, the U.S. Office of Personnel Management, Office of Diversity and Inclusion's *Government-Wide Diversity and Inclusion Strategic Plan 2011.* That document defines diversity broadly, including but not limiting it to "national origin, language, race, color, disability, ethnicity, gender, age, religion, sexual orientation, gender identity, socioeconomic status, veteran status, and family structures" (p. 5), as well as differences in place of origin, experiences, and thoughts. This plan asserts a business case for increasing workforce diversity and that individuals with "varying degree types," specifically those from "Science, Technology, Engineering and Mathematics (STEM) backgrounds . . . will also benefit agencies and offices Government-Wide" (p. 4). The plan calls for new metrics, along with new human resource practices, to ensure that STEM specialists are hired and promoted to ensure the diversity of the federal workforce. As with previous definitions of diversity, the one given here is also too broad for DoD to feasibly use in goal-setting and performance assessment.

DoD's Current Diversity Strategic Plan Addresses STEM as Part of Diversity

Executive Order 13583 calls for each federal agency to develop and implement a strategic diversity and inclusion plan. To comply with the order, DoD developed and released the *Diversity and Inclusion Strategic Plan, 2012–2017* in early 2012. The plan emphasizes why DoD considers diversity to be imperative to the success of the organization as well as the nation in its introduction:

> We gain a strategic advantage by leveraging the diversity of all members and creating an inclusive environment in which each member is valued and encouraged to provide ideas critical to innovation, optimization, and organizational mission success. (DoD, 2012, p. 3)

This passage highlights the way in which this plan, like the executive order that it supports, has a foundation based on the assumption of a "business case" for diversity. The business case for diversity regards personal differences—demographic, cognitive, structural, and national—as capabilities that, when managed properly, can achieve desired goals and outcomes (see MLDC, 2011). DoD's diversity and inclusion strategic plan asserts that diversity can increase organizational effectiveness, performance, and agility and stresses that the inclusive environment it advocates for is based on the principles of equal employment opportunity (EEO) and military equal opportunity (MEO). The plan also stresses the importance of DoD leveraging the talents of men and women with different backgrounds to accomplish its defense and humanitarian missions, and that the differences between people can improve mission effectiveness and innovative capability.

Three goals and supporting objectives, strategic actions, and initiatives presented in the *Diversity and Inclusion Strategic Plan, 2012–2017* are intended to provide direction for DoD leaders to "create . . . a more diverse talent pool for DoD military accessions and civilian hires" (DoD, 2012, p. 4). We summarize the goals and objectives below; the full document is available to the public online.

- Goal 1: Ensure leadership commitment to an accountable and sustained diversity effort: A strong commitment to the new diversity vision is essential for success; in this vein, the first goal is to generate the support, accountability, and capability of top leadership throughout DoD. Two objectives are offered for Goal 1. The first is to align leadership with current policies and make them accountable through new accountability metrics. The second is to arm leaders with communication tools and messages that proactively inform internal and external audiences about DoD's diversity workforce goals.
- Goal 2: Employ an aligned strategic outreach effort to identify, attract, and recruit from a broad talent pool: This goal aims to build a diverse talent pool of qualified candidates from varied backgrounds who have experience and education in areas related to DoD's many mission-critical occupations. There are two objectives associated with Goal 2. Together, both goals recommend ways that DoD can assess, synchronize efforts, and establish new ways to reach out to all segments of society. Of particular relevance to DoD STEM leaders is a recommendation calling for the establishment or expansion of "strategic relationships with inter-

nal and external key stakeholders at diverse colleges and universities, trade schools, apprentice programs, Science, Technology, Engineering, and Mathematics (STEM) initiative programs, and affinity organizations" (DoD, 2012, p. 8).

- Goal 3: Develop, mentor, and retain top talent from across the Total Force: Executive Order 13583 instructs leaders of the Armed Services and federal agencies to make changes in the ways in which their workforces are recruited and hired as well as promoted and retained. It urges leaders to begin to make a concerted, organized effort to "create a culture that encourages collaboration, flexibility, and fairness to enable individuals to participate to their full potential" (DoD, 2012, p. 15). Toward this end, the objectives under this goal ask leaders to consider programs, practices, and policies that support professional and personal development of the workforce.

Summary

In our review of national and DoD policy, we have documented overlapping STEM and diversity workforce goals that ASD(R&E) and ODMEO may consider as they continue to develop studies, programs, and further policies to improve the diversity of the DoD STEM workforce. Current legislation and STEM- and diversity-related strategic plans assert they are based on three factors of concern to policymakers:

- There is a need for talented and innovative STEM workers to meet 21st century global challenges.
- The nation's demographic makeup is changing and will continue to change.
- The federal workforce, including DoD, must be inclusive and reflect the demographics of the country it serves.

While these policy documents assert the importance of creating a diverse federal and DoD workforce, their definitions of diversity are far too broad to use when evaluating how well the DoD is currently performing overall or in specific programs, or in assessing the likely efficacy of possible future changes. A next step for DoD should be formally articulating which aspects of diversity it wishes to focus on for its workforce, as well as specific goals.

The next chapter will describe demographic trends and the DoD STEM workforce. We show that as the country becomes increasingly Hispanic, if educational and STEM occupational rates for each racial/ethnic group do not converge, STEM workers will become increasingly nonrepresentative of the country as a whole. This decline may make it even more difficult for DoD to attain its goal of a more racially/ethnically representative STEM workforce.

CHAPTER THREE

Demographic Trends and the DoD STEM Workforce

The demographic makeup of the country is changing. To be inclusive, we need to take advantage of underserved communities. (Reginald Brothers, DoD STEM Diversity Summit, November 2012)

DoD has made it a priority to make its workforce more representative of the country as a whole. At the Summit, the most-discussed aspect of national diversity was racial/ethnic differences. As we demonstrate in this chapter, the nation is changing demographically. Overall, the U.S. population is increasingly more Hispanic and less non-Hispanic white. Meanwhile, the makeup of the U.S. STEM workforce (from which DoD draws its STEM employees) is significantly more white and Asian than the overall working-age population. Given the significantly lower representation of Hispanics in the STEM workforce and among STEM college graduates, this demographic shift presents a major challenge to DoD's efforts to make its STEM workforce more representative of the nation. Without significant increases in STEM participation on the part of Hispanics, or decreases on the part of whites or Asians, the gap between the proportion of Hispanics in the working-age population and among STEM workers will grow significantly. While we have no reason to believe that current rates of STEM participation will remain the same among each racial/ethnic group, we present this as a very preliminary analysis of what the results of that would be. We also provide analysis showing that current STEM racial/ethnic differences are being driven in large part by differences in educational attainment.

We make no broad claims here about the inadequacy of the current or future STEM workforce to meet demand, maintain the United States' prominent role in the STEM fields, or maintain national security, or about the need for STEM fields to diversify in order to achieve any of these objectives. These assertions can be found in other sources, including the 2011 National Academies report *Expanding Underrepresented Minority Participation: America's Science and Technology Talent at the Crossroads.* Counterarguments to these assertions have been offered as well, and can be found in sources including the 2012 National Academies report *Assuring the U.S. Department of Defense a Strong Science, Technology, Engineering, and Mathematics (STEM) Workforce* and the RAND monograph *U.S. Competitiveness in Science and Technology* (Galama and Hosek, 2007). Evaluating these claims is beyond the scope of this report. Here, we focus only on helping DoD meet its diversity goals in part by understanding possible future challenges associated with changing American demographics.

The definition of STEM occupations used in this report comes from the U.S. Department of Commerce report *STEM: Good Jobs Now and for the Future* (Langdon et al., 2011), combined with the Bureau of Labor Statistics crosswalk information between standard occupational classification codes and census categories in *2010 Census Occupational Classification: Major Occupational Groups and Detailed Occupations Used in the Current Population Survey Beginning January 2011.* The broad occupational categories classified as STEM by these sources are computer and mathematical occupations; engineering and surveying occupations; life, physical, and social science occupations; and STEM managerial occupations. This list does not include health practitioners, such as nurses and doctors. However, because the ODMEO definition of STEM workers does include health practitioners, Figure 3.5, which compares the DoD STEM workforce with the overall civilian STEM workforce, includes health practitioners in its analysis of the overall STEM workforce. Based on examining other definitions of STEM occupations, including in Georgetown University's Center on Education and the Workforce's *STEM* (Carnevale, Smith, and Melton, 2011) and the Bureau of Labor Statistics' "Science, Technology, Engineering, and Mathematics (STEM) Occupations: A Visual Essay" (Cover,

Jones, and Watson, 2011), we found DoD's inclusion of health workers in STEM occupations to be somewhat nonstandard.

The definition of STEM academic fields used comes from the Department of Commerce paper *Women in STEM: A Gender Gap to Innovation* (Beede et al., 2011), which includes computer, math, engineering, and physical and life sciences. It excludes social science majors, psychology, and medical fields, such as nursing and pre-med. Our degree analysis includes only bachelor's degrees, excluding both associate's degrees and graduate degrees.

Demographic Shifts Combined with Current STEM Rates Would Create Bigger Gaps Between the Overall National and STEM Workforces

The Census Bureau's 2008 National Population Projections demonstrate that the United States is in the midst of a major demographic shift toward becoming significantly more Hispanic and less white, non-Hispanic.[1] Figure 3.1 graphically illustrates this shift: The Census Department estimates that in 2000, whites not of Hispanic origin (hereafter referred to as whites) made up about 70 percent of 18–65-year-olds, and projects that in 2050 they will make up less than 45 percent, with Hispanics increasing from 12 to 30 percent. The proportion of Asians is also projected to increase, from about 4 percent in 2000 to about 8 in 2050.[2] These projections rely on various assumptions about mortality rates, birth rates, and migration.

---

[1]   All of the projections data in this chapter on the current and future racial/ethnic makeup of the country are from the 2008 National Population Projections (U.S. Census Bureau, 2008). All of the data on current U.S. educational attainment, STEM degree attainment, and likelihood of working in a STEM occupation are from the 2010 American Community Survey (ACS; Ruggles et al., 2010). Projections having to do with future educational attainment, STEM degree attainment, and likelihood of working in a STEM occupation are from combining 2010 ACS propensities for those outcomes by racial/ethnic group with the 2008 National Population Projections data.

[2]   The "other" group, in these data as well as the ACS data shown later in this chapter, includes both non-Hispanics who are multiracial and non-Hispanics who are American Indian, Alaskan Natives, or Native Hawaiian or other Pacific Islander.

Figure 3.1
Census Projections, 2000–2050, 18–65-Year-Olds

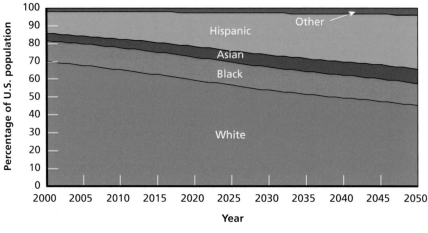

SOURCE: Based on authors' computations with 2008 Census projections data (U.S. Census Bureau, 2008).
RAND RR329-3.1

Our findings on STEM and demographics are consistent with the National Academies' report *Expanding Underrepresented Minority Participation: America's Science and Technology Talent at the Crossroads*: We find that the U.S. STEM workforce is "overwhelmingly white and Asian." Figure 3.2 shows the current composition of the 23–29-year-olds, the proportion of those who have college degrees and work in STEM occupations, and the proportion who have STEM degrees.[3] Hispanics make up 20 percent of the overall young adult population, but only 5 percent of those with a college degree who work in STEM occupations, and 7 percent of those with degrees in STEM fields.

---

[3]   In general, we look at 23–29-year-olds in this report because there are different age distributions in this country for each racial/ethnic group, and we did not want to confuse differences due to age with those due to race/ethnicity. We believe the current education and occupational propensities of young adults are the most relevant to future cohorts, and looking at the behavior of older adults would be less significant.

Figure 3.2
2010 Percentage of 23–29-Year-Olds in Overall Population, Among STEM Degree Holders, and Among Those with College Degrees in STEM Occupations, by Race/Ethnicity

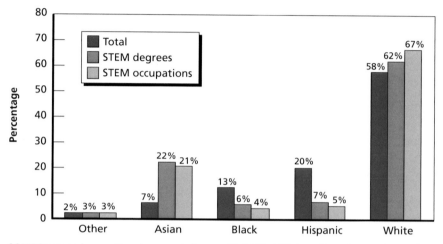

SOURCE: Based on authors' computations with 2010 ACS data (Ruggles et al., 2010).
RAND *RR329-3.2*

Figure 3.3 shows the projected numbers for 2050, if current rates of STEM degree attainment and STEM occupations among young college graduates stay the same for each racial/ethnic group. We do not anticipate this will be the case: The proportion of each group pursuing STEM degrees and employment is not static. However, because it is beyond the scope of this study to predict how these proportions will change over the coming decades, we present these results as a very preliminary thought experiment. Under this assumption, Hispanics will make up a greater percentage of young people with STEM degrees and young people with college degrees who work in STEM occupations than they currently do now. However, *the gap between the Hispanic proportion in the population as a whole and the Hispanic proportion in STEM areas will also increase.* For instance, Hispanics currently make up 20 percent of 23–29-year-olds and 7 percent of 23–29-year-olds with STEM degrees. If the propensity to obtain a STEM degree holds, in 2050 Hispanics will make up 35 percent of 23–29-year-olds but

Figure 3.3
Projected 2050 Percentage of 23–29-Year-Olds in Overall Population, Among STEM Degree Holders, and Among Those with College Degrees in STEM Occupations, by Race/Ethnicity

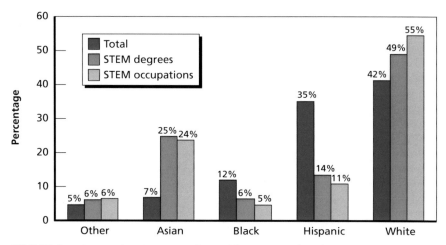

SOURCE: Based on authors' computations with 2010 ACS data (Ruggles et al., 2010) and 2008 Census projections data (U.S. Census Bureau, 2008).
RAND RR329-3.3

only 14 percent of 23–29-year-olds with STEM degrees. This gap will make recruiting and retaining a nationally representative DoD STEM workforce more challenging.

Women are also significantly underrepresented in STEM fields relative to their proportion in the overall population. Figure 3.4 shows that, while women make up half of 23–29-year-olds, they make up only 40 percent of college graduates with STEM degrees and 31 percent of college graduates working in STEM occupations. Since the proportion of women in the population is fairly constant, we do not project an increasing gap between the proportion of women in the population and the proportion in the STEM workforce. While the emphasis of this report is on racial/ethnic diversity, some interest in gender diversity was expressed at the Summit as well, so these data may also be of interest to DoD.

Figure 3.4
2010 Percentage of 23–29-Year-Olds in the Overall Population, Among STEM Degree Holders, and Among Those with College Degrees in STEM Occupations, by Gender

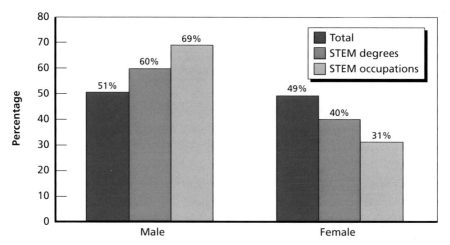

SOURCE: Based on authors' computations with 2010 ACS data (Ruggles et al., 2010).
RAND *RR329-3.4*

DoD STEM Workers Are as Racially/Ethnically Diverse as STEM Citizen Workers Overall

The racial/ethnic makeup of the DoD civilian STEM workforce closely resembles the makeup of citizens employed in STEM occupations in the United States, which is much more white than the population as a whole. ODMEO provided us with information on the composition of the DoD civilian STEM workforce as of the end of FY 2012. Because ODMEO included health care practitioners, we included those in STEM workers as well, just for this figure.

Figure 3.5 shows the ODMEO numbers side-by-side with the makeup of citizens employed in STEM occupations in the United States from 2010 ACS data.

As shown, the DoD civilian STEM workforce has a slightly higher black population and less Asian representation. There is also slightly greater representation of "other" groups than the overall STEM citizen workforce, which could be due to reporting differences, such

Figure 3.5
Overall STEM Citizen Workforce and DoD Civilian STEM Workforce, by Race/Ethnicity and Gender

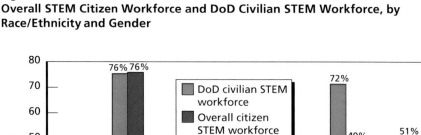

SOURCE: Based on authors' computations with 2010 ACS data (Ruggles et al., 2010) and with DoD-provided data.
NOTE: There are two major differences between the ODMEO STEM occupation definition and the one we use throughout this report: The ODMEO definition includes health practitioners and program management occupations. As a result, for this figure and only this figure, we included health practitioners in our definition of STEM workers. This is as close as we can get to bringing our definition of STEM occupations in line with that of ODMEO, since program management is not a category in the Census data we used.
RAND RR329-3.5

as people who choose not to disclose racial information being classified as "other." Generally, though, the DoD numbers closely parallel the overall STEM labor market in terms of racial/ethnic composition. This similarity is a good indication that whatever factors affecting the composition of the overall STEM workforce also affect DoD STEM hiring. It also suggests there are major barriers to the DoD reaching its goal of a STEM workforce that is significantly more representative of the country.

The overall STEM workforce, using the DoD definition, which includes health practitioners, is more than half female, whereas the DoD civilian STEM workforce is only 29 percent female. We attribute this to the fact that health practitioners make up over half of citizen

STEM workers, and women make up a large majority of that group. In contrast, we believe that health practitioners constitute much less than half of the DoD STEM workforce, although we do not have data to support this speculation. Excluding health care workers had only a small effect on the racial/ethnic composition of the citizen STEM workforce, but it caused the gender composition to be very similar to the DoD STEM workforce.

Current STEM Gaps Are Driven Partly by Educational Differences

Current racial/ethnic differences in STEM degree attainment and STEM occupations are driven partly by differences in educational attainment. Asian and white young adults graduate from high school and college at higher rates than black and Hispanic young adults. Figure 3.6 shows these differences. Ninety-one percent of Asian young adults age 23–29 are high school graduates, compared with 89 percent of whites, 81 percent of blacks, and 67 percent of Hispanics. General Education Development Test certificate (GED) attainment or other alternative high school certification is not classified here as high school graduation. College graduation percentages follow the same pattern: 49 percent of Asian young adults have completed a bachelor's degree or higher, compared with 35 percent of whites, 17 percent of blacks, and 12 percent of Hispanics.

Asians and whites are also more likely than blacks and Hispanics to hold undergraduate degrees in STEM fields.[4] Most of these differences are driven by educational attainment rates. In other words, because Asians and whites are more likely to graduate from high school as well as from college, they are also more likely to have college degrees in STEM fields. A smaller part of the gap is due to differences in majors among college graduates. Figure 3.6 shows that 22 percent of Asian young adults hold bachelor's degrees in STEM fields, compared

---

[4]   We define STEM fields as those related to math, engineering, computers, and physical and life sciences. We exclude medical fields, psychology, and the social sciences.

Figure 3.6
Rates of Educational Attainment Among 23–29-Year-Olds, by Race/Ethnicity and Gender

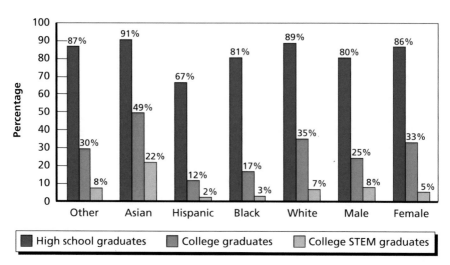

SOURCE: Based on authors' computations with 2010 ACS data (Ruggles et al., 2010).
RAND RR329-3.6

with 7 percent of whites, 3 percent of blacks, and 2 percent of Hispanics. However, as Figure 3.7 shows, there are no significant differences between rates of STEM majors among young white, black, and Hispanic college graduates.[5] The exception to degree differences being driven by overall college attainment rates is that Asian college graduates are much more likely to hold STEM degrees than college graduates from any other group: 45 percent of Asian college graduates have STEM majors, more than twice that of whites, blacks, or Hispanics.

In contrast, differences between the proportions of men and women with STEM degrees are driven solely by differences in majors between male and female college graduates, since women have both higher educational attainment rates and lower STEM attainment rates.

---

[5]   For instance, Figure 3.6 shows that 35 percent of young white adults are college graduates, while Figure 3.7 showing that 20 percent of young white college graduates have STEM degrees. Therefore, 7 percent of young white adults have undergraduate degrees in STEM fields, which is the percentage shown in Figure 3.6.

Figure 3.6 shows that 33 percent of young adult women are college graduates, compared with 25 percent of young adult men, but that 8 percent of young men hold college degrees in STEM fields, compared with only 5 percent of women. And as Figure 3.7 shows, male college graduates are about twice as likely as female college graduates to hold a degree in a STEM field. The STEM degree gender gap, unlike the STEM degree racial/ethnic gap, is not a product of differences in educational attainment, but of differences in choice of degree among male and female college graduates.

Differences in Rate of College Graduate STEM Workers

Asians are significantly more likely to be college graduates employed in STEM occupations, followed by whites, blacks, and Hispanics, as seen in Figure 3.8. Nine percent of Asian young adults are college graduates who are employed in STEM occupations, compared with 3 percent of whites, 1 percent of Hispanics and blacks, and 3 percent of other young adults.

Figure 3.7
STEM Degree Attainment Among 23–29-Year-Old College Graduates, by Race/Ethnicity and Gender

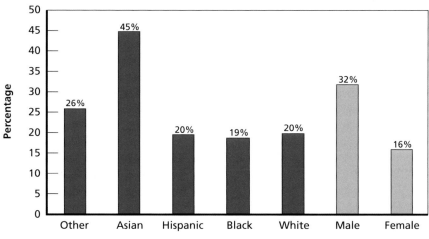

SOURCE: Based on authors' computations with 2010 ACS data (Ruggles et al., 2010).
RAND RR329-3.7

Figure 3.8
Percentage of 23–29-Year-Olds Who Are College-Educated STEM Workers,
by Race/Ethnicity and Gender

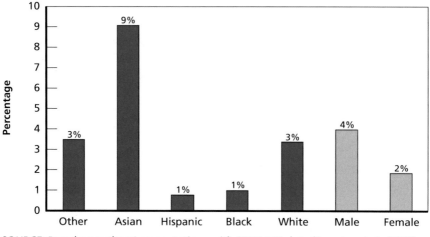

SOURCE: Based on authors' computations with 2010 ACS data (Ruggles et al., 2010).
RAND RR329-3.8

However, as with the differences in STEM degrees, much of the difference in the proportion of young adults of each racial/ethnic group who have college degrees and work in STEM jobs by group is being driven by differences in educational attainment rather than differences between college graduates. Figure 3.9 shows the differences in the proportion of STEM workers among young college graduates. About 18 percent of Asian college graduates are employed in STEM occupations, compared with 7 percent among Hispanics, 6 percent among blacks, and 10 percent among whites. Among whites, blacks, and Hispanics, differences in educational attainment are driving differences in the rates of college STEM workers much more so than differences between types of jobs worked by college graduates in each group. For instance, Figure 3.6 shows that Hispanic and black young adults are college graduates at less than half the rate of white young adults. In contrast, the gaps between the proportions of college graduates in each group who are working in STEM jobs are smaller, as shown in Figure 3.9. The exception, again, is Asians, who are substantially more likely both to be college graduates and, if they are college graduates, to work in STEM jobs.

However, the difference in the proportion of men and women who are college graduates working STEM jobs is driven solely by male college graduates being much more likely to work in STEM jobs, rather than by differences in overall degree attainment. Figure 3.8 shows that 4 percent of young men are college graduates in STEM jobs, but only 2 percent of young women. As Figure 3.9 shows, this is because male college graduates are more than twice as likely to work in STEM jobs as female college graduates. This is related to the lower propensity of female college graduates to have majored in a STEM field, since STEM majors are more likely than non-STEM majors to go on to work in STEM fields after college.

Figure 3.9
Percentage of 23–29-Year-Old College Graduates Who Are STEM Workers, by Race/Ethnicity and Gender

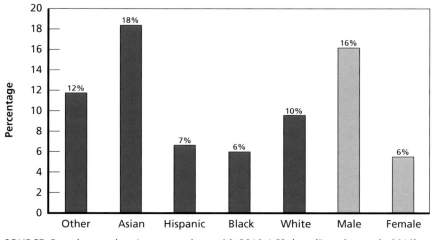

SOURCE: Based on authors' computations with 2010 ACS data (Ruggles et al., 2010).
RAND RR329-3.9

Summary

As the United States' overall demographics change, achieving the goal of having a STEM workforce that mirrors the population will become increasingly difficult for DoD without a significant increase in the proportion of Hispanics who enter STEM fields. The differences in the proportion of college graduates in STEM majors among white, black, and Hispanic workers are due solely to differences in educational attainment, whereas the differences in the proportions of each group that hold a college degree and work in a STEM occupation are due both to differences in college degree attainment and racial differences in likelihood of working in a STEM occupations among college graduates. Without significant racial convergence in STEM participation, the gap between the proportion of Hispanics in the working-age population and among STEM workers will grow significantly. In contrast, gender differences in STEM outcomes are not driven by differences in overall educational attainment, since women are more likely to hold college degrees, and demographic changes will not necessarily impact the gender distribution in STEM fields.

CHAPTER FOUR

Current DoD STEM Initiatives

> We look at our STEM investment not just as a K–12 investment, but also as a continuum [for] broadening the talent pool (Jeffery Singleton, Director for Basic Research, Department of the Army, DoD STEM Diversity Summit, November 2012).

Over the course of the Summit, leaders from the Armed Forces presented descriptions of programs designed specifically to develop qualified STEM professionals from the U.S. population. In this chapter, we offer some initial observations and analysis drawn from the presentations and the information made available to us after the Summit. We do not include all programs discussed or presented at the Summit; for example, the programs developed or sponsored by Historically Black Colleges and Universities (HBCUs) that were discussed are not represented, as we limit our focus to those programs for which DoD has primary responsibility. We summarize DoD component and National Guard presentations in Appendix B.

Program Information Collection and Limitations

Table 4.1 at the end of this chapter sumarizes the 58 outreach programs described by DoD component leaders over the course of two days, and is organized according to presentation order, with the Navy presenting first. After the name of the program, we present the following information:

1. the program's intended participants
2. an overview of the program

3. a percentage reflecting the number of minority participants
4. the number of total participants annually
5. a brief program description
6. how program leaders currently assess the program
7. the program's annual cost.

Information for Table 4.1 was taken directly from the Summit presentations when available; the research team asked presenters for more information on their programs after the Summit, but not all presenters responded to queries, or responded in time to be included in this document. In some cases, such as statistics related to minority participation for certain programs, no information is collected, as demographic diversity may not be a formally recognized goal. Notably, the kinds of information available at this time vary from program to program. For example, we requested numbers of annual participants of programs; in some cases, the only information available was the total number of participants since the program's inception.[1]

The incomplete nature of the information presented in Table 4.1 indicates this effort may be the first to garner a comprehensive view.

DoD STEM Outreach Efforts Have Different Goals

As Table 4.1 suggests, program goals vary greatly. They range from broadening youths' interest in science to meeting STEM hiring goals. Some of the programs state explicit goals related to demographic diversity, while others do not. Program goals include providing academic enrichment to poor, rural, or underrepresented minority children; improving academic performance among the children of service members in DoD-supported schools; recruiting students from in-demand STEM fields; and advancing military research.

Many of the programs go significantly beyond the scope suggested in the 2012 National Academies' report *Assuring U.S. Depart-*

---

[1] The content of Table 4.1 is taken directly from materials given to us by the services, with spelling and grammatical edits made in the interests of readability, but otherwise unchanged.

ment of Defense a Strong Science, Technology, Engineering, and Mathematics (STEM) Workforce. The report argued that because the DoD STEM workforce is such a small percentage of the overall workforce, DoD should focus not on trying to change the overall STEM workforce. Instead, DoD should put effort into fulfilling its own recruiting needs. Many of the current DoD STEM programs have goals that fall outside of that target. For instance, academic enrichment programs targeted toward K–12 students that have no military element (as JROTC does) are unlikely to significantly increase the probability that participants will someday work as STEM professionals within the DoD civilian workforce.

However, these programs are very much in line with recommendations suggested in the National Academies' 2011 report *Expanding Underrepresented Minority Participation: America's Science and Technology Talent at the Crossroads.* Here it is argued that increasing minority STEM participation needs to be an urgent national priority and that programs aimed at all levels of education from various stakeholders are necessary to accomplish this goal. Notably, the National Academies' 2012 report especially praises the Science, Mathematics, And Research for Transformation (SMART) Scholarship program, recommending that DoD expand SMART and programs like it. The report mentions the high academic qualifications of its students, calling it "well-targeted to the nation's national security needs" (National Academies, 2012, p. 6). SMART fits into the report's goals of improving the quality of the DoD civilian workforce without regard to its demographic makeup.

In addition to a variation in goals, there seems to be significant variation in the size of programs (in terms of both number of participants and costs), whether the programs are centralized and national or locally run and localized, the degree of military involvement (running its own programs versus financially sponsoring programs run by external organizations), and the degree of assessment being performed.

Metrics and Assessment Needed to Quantify and Qualify Success

Currently, program assessment varies significantly across the services and for each program. Presenters at the Summit did not offer information about existing assessment for many of the programs they described. Others described assessments for K–12 programs that collected information on outcomes not directly related to student achievement or participation in STEM fields. For example, facilitators of a middle school student program measured whether the program affected participants' interest in studying engineering, as well as how it affected their problem-solving confidence. While interest in studying engineering is likely higher among students who go on to study engineering than students who do not, it does not necessarily follow that a program raising engineering interest will increase actual engineering participation. Assessment of advanced students, when we found record of it, tends to focus on outcomes that are of more direct interest. For example, facilitators of a college student program measured how academic achievement and enrollment in graduate school changed as a result of participating in the program.

Still Missing Much Information

In addition to clarifying what goals it wishes to pursue via these programs, DoD would benefit from standardizing program information and expanding and standardizing assessment. The Navy provided significant information on service-specific programs, including target audience, scope, funding, and assessment. Other services provided less information and may be less aware of the details of the STEM and/ or diversity programs they are funding, particularly in cases where programs are locally run as opposed to centrally. For most programs described over the course of and after the Summit, we were not told about assessment. We also were not informed about other elements critical to effective program design, such as goals and cost. We note that many programs do not measure minority participation, which of course was a special concern to Summit participants.

Summary

This chapter documents some of DoD's many outreach efforts designed to generate national interest in STEM as well as provide STEM education. We observe that the goals and intended participants of these efforts vary greatly. Many efforts seek to provide educational support, such as SeaPerch, the Navy's robotics competition, which is designed to increase school-age children's interest in STEM and problem-solving skills. Other educational programs, such as the Air Force's Teachers Materials Camp, seek to support school curriculum development. Several programs directly "feed" into the DoD workforce.

This initial review of the program goals, participants, and existence of diversity goals suggests that a full assessment of programs is needed, which is especially true if DoD leaders choose to pursue the goal of producing a steady flow of qualified STEM professionals representing the nation demographically. It is uncertain how well each program is accomplishing that goal, but by working more closely together and improving assessment, DoD will be able to more efficiently use its limited resources to pursue its goals.

Table 4.1
List of Current STEM/Diversity Programs

Service	Name	Level of Student/ Employee	Overview	Minority Participation	Number of Participants	Description	Assessment	Annual Cost
Navy	SeaPerch	Middle school	Middle school robotics competition	45%	35,000 students, 4,000 teachers *missing number of annual participants (total, not annual)*	"Gateway" robotics program for middle and high school featuring a submersible remotely operated vehicle (ROV)	Participating in SeaPerch increased interest in studying engineering in 25% of middle school and 30% of high school students	$1,400,000 from ONR $750,000 from SYSCOMS and NDEP *(total, not annual)*
						Through robotics and underwater ROVs, students learn about careers in naval architecture, and marine, ocean, and naval engineering	Program improved problem solving confidence in 34% of middle school and 43% of high school students	

Table 4.1—Continued

Service	Name	Level of Student/ Employee	Overview	Minority Participation	Number of Participants	Description	Assessment	Annual Cost
Navy	Iridescent Technovation Challenge	High school	Girls app-development competition	40%	730	Entrepreneurial team competition for App development for young women in high school Teams pitch their App and business plan to panel of venture capitalists Winning App is professionally developed and released Each team paired with a female graduate or undergraduate student near-peer mentor Partnership with Google, Microsoft, LinkedIn, Twitter, MIT, Stanford, Berkeley, UCSF	Longitudinal study of program impact will measure: • Student commitment to STEM education • Student attitude toward, pursuit of, and involvement in STEM/ STEM career • Student increase in STEM concepts and content • Parent/family awareness and interest in STEM and STEM careers • Longitudinal student tracking for continued participation in STEM programs	$850,000 annually

Table 4.1—Continued

Service	Name	Level of Student/ Employee	Overview	Minority Participation	Number of Participants	Description	Assessment	Annual Cost
Navy	Iridescent Family Science Program	Elementary and middle school	Hands-on after-school program	95%	7,270	A hands-on, experiential learning after school program Customized 10 Navy-relevant learning modules Leverages near-peer mentors from USC, NYU-Poly, Cooper Union Aimed at 3rd-7th grade underserved, underprivileged children and their families Naval veterans involved through The Mission Continues program	After participating in program, 80% of students interested in pursuing STEM education/career	$1,500,000 annually
Navy	National Math and Science Initiative	High school	AP courses for military dependents	26%	800	Part of Initiative for Military Families and First Lady's Joining Forces Initiative Providing proven AP STEM curriculum to high school with high percentages of military dependents	AP math and science passing scores increased by 57% (7 times greater than the national average) Students passing AP exam are 3 times more likely to earn a college degree	$375,000 annually

Table 4.1—Continued

Service	Name	Level of Student/ Employee	Overview	Minority Participation	Number of Participants	Description	Assessment	Annual Cost
Navy	SEAP	High school	Internship program	21%	215			
Navy	Sally Ride Science and ASM Teacher Training	Middle and high school	Teacher training programs	From rural Alabama, Mississippi, and Louisiana	200			
Navy	Summer Camps (Crime Scene Investigation, National Society of Black Engineers)	Middle school	Hands-on camps	80%	300			
Navy	NREIP	College	Internship program at the labs and centers	15%	155			

Table 4.1—Continued

Service	Name	Level of Student/ Employee	Overview	Minority Participation	Number of Participants	Description	Assessment	Annual Cost
Navy	Naval Research Laboratory STEM Academy	College	Minority institution–focused college internships at National Research Laboratory	100%	45	Undergraduate and graduate students from HBCUs and MSIs Hands-on, experiential research internships at Labs and Warfare Centers alongside scientists and engineers Interns exposed to larger naval S&T community through seminars, tours, and field trips	Planned Metrics and Assessment: • Demographics of applicants and interns • Selectivity • Returning as interns or participate in other Naval STEM programs • Pursuing STEM Education/ Degree	$330,000 FY11; $730,000 anticipated FY12
Navy	Florida International University	College	Reinventing curriculum for basic STEM courses	83%	Development beginning in fall			

Table 4.1—Continued

Service	Name	Level of Student/ Employee	Overview	Minority Participation	Number of Participants	Description	Assessment	Annual Cost
Navy	UT Pan American	College	Developing 10–15 Navy-relevant STEM courses	97%	1,700	Collaboration between 5 Hispanic-serving institutions in South Texas Center will support professional development for faculty to create 10–15 Navy-relevant STEM courses Center will support undergraduate research in Navy-relevant area Faculty will develop and standardize curriculum for Texas pre-freshman STEM outreach program	Ongoing metrics and assessment demographics: • Track student retention in STEM • Graduation with STEM degree • Enrollment in graduate school • Employment by Navy/DoD labs • Academic achievement (GPA) • Tracking of all publications, presentations, and patents resulting from student participation • Tracking of all fellowships, scholarships, and awards received by student participants	$1,000,000 Annually (up to 4 years) from OSD DoD HBCU/MI Education and Research Funds

Table 4.1—Continued

Service	Name	Level of Student/ Employee	Overview	Minority Participation	Number of Participants	Description	Assessment	Annual Cost
Navy	Business–Higher Education Forum Higher Ed STEM Model	College	Developed model of best practices for higher education retention programs	Launch fall 2012	To be used to select future naval programs			
Navy	Digital Tutor Grand Challenge	Middle School	Development of middle school and new recruit STEM tutor	October 1 start date	4 awards			
Navy	Gooru	5th–12th grade	Online student and teacher resource	60%	4,500 students/ 200 teachers			

Table 4.1—Continued

Service	Name	Level of Student/ Employee	Overview	Minority Participation	Number of Participants	Description	Assessment	Annual Cost
Navy	Youth Exploring Science Program	High school	Academic support and life skills to high school students	90%	246	4-year high school program for St. Louis area teenagers ages 14–18 Partnership with St. Louis Science Center Focus on minorities, disadvantaged and at-risk students Provides academic support and life skills development in a work-based, inquiry-learning science environment Creating roadmap for program distribution to other Science Centers	Ongoing metrics and assessment: • High school graduation • College enrollment • Career choice • External evaluation funded by ONR • Track participants over 4 years • Evaluate program impact on college and career choices • Evaluate understanding of STEM concepts and content • Track student participation in other STEM activities	$580,000 annually

Table 4.1—Continued

Service	Name	Level of Student/ Employee	Overview	Minority Participation	Number of Participants	Description	Assessment	Annual Cost
Navy	HBCU Tuskegee University MS Systems Engineering Program	Graduate school	Graduate program with work requirement	100%	12	Student awarded 1-year scholarship for a masters in science in systems engineering degree with a 3-year work commitment at Naval Sea Systems Command Students exposed to a highly tailored master's of science systems engineering curriculum developed by the Naval Postgraduate School Emphasizes Navy-relevant technologies Enhances Naval Lab workforce diversity through active engagement with HBCU/minority institution students and faculty	33 graduates, now full-time employees Employed by 6 Warfare Centers 97% completion rate (1 loss)	$1,200,000 from Section 852 Funds $600,000 (half from NAVSEA)

Table 4.1—Continued

Service	Name	Level of Student/ Employee	Overview	Minority Participation	Number of Participants	Description	Assessment	Annual Cost
OSD, Reserve Affairs	Starbase	5th grade	Extracurricular STEM program for 5th graders	Minorities are one of the target populations	75,000 students annually	School-based (after school program)	Pre/post-test show gain of 6.34 points Attitudes of all participants shift dramatically toward the positive Youth leave with feeling of empowerment Commanders have positive perception of the program	Academy cost: $330,000 Average cost per student: $306
OSD, Reserve Affairs	Starbase 2.0	6th–12th grade	STEM-related mentoring and clubs	Minorities are one of the target populations	230 mentors and 700 students in 2012	Team mentoring targeting middle school students who have participated in STARBASE Clubs meet 4 hours per month for 6–9 months; 3–4 mentors, 10–15 students per club STEM activity-based mentors with STEM careers or strong STEM interest		

Table 4.1—Continued

Service	Name	Level of Student/ Employee	Overview	Minority Participation	Number of Participants	Description	Assessment	Annual Cost
Air Force	Awards to Stimulate & Support Under-graduate Research Experiences (ASSURE)	College	Under-graduate research support?					$4.5 million for FY13
Air Force	National Defense Science & Engineering Graduate (NDSEG) Fellowship Program	Graduate school	Graduate fellowship					$38 million for FY13
Air Force	University NanoSatellite Program	College?						$1.6 million for FY13
Air Force	Air Force Research Laboratory Section 219 funding for STEM outreach	Hiring?						$6.1 million for FY13
Air Force	K–12 National Defense Education Program (NDEP)	K–12						

Table 4.1—Continued

Service	Name	Level of Student/ Employee	Overview	Minority Participation	Number of Participants	Description	Assessment	Annual Cost
Air Force	Teachers Materials Camp	K–12?	Teacher training programs	Targets high school teachers from under-represented communities		Targets high school teachers from under-represented communities Leverages academia/ universities who recruit teachers for camp Joint effort with AF STEM Outreach Coordination Office and Air Force Diversity Office Provides Air Force scientists and engineers as guest speakers from same under-represented communities Connected 6 Air Force installations with established Materials Camp program started by the Department of Education and National Science Foundation and NSF—5 new camps established		

Table 4.1—Continued

Service	Name	Level of Student/ Employee	Overview	Minority Participation	Number of Participants	Description	Assessment	Annual Cost
Army	University Based Research	College, graduate school, researchers	Army-sponsored research		The University Single Investigator Program consists of about 1,200 grants to about 340 universities	Consists of University-Affiliated Research Centers, Centers of Excellence for Enduring Army Needs, Multidisciplinary University Research Initiatives, and University Single Investigator Program. These are all programs in which military-funded research is done at universities.		

Table 4.1—Continued

Service	Name	Level of Student/ Employee	Overview	Minority Participation	Number of Participants	Description	Assessment	Annual Cost
Army	Collaborative Technology / Research Alliances	College, graduate school, researchers	Army-sponsored research			These are all programs in which military-funded research is done at multiple universities Projects are on robotics, cognition and neuroergonomics, micro autonomous systems technology, network science, network and information science, international technology alliance, environments, materials in extreme dynamic environments, and multiscale multidisciplinary modeling of electronic materials		

Table 4.1—Continued

Service	Name	Level of Student/Employee	Overview	Minority Participation	Number of Participants	Description	Assessment	Annual Cost
Army	HBCU/MI Partnerships in Research Transitions	College, graduate school, researchers	Army-sponsored research with diversity goal	Partners are minority institutions		Partners are Howard, Hampton, NC A&T, Delaware State Provides for and supports a wide diversity of scientific research and idea generation Supports basic research through the Partnership in Research Transition (PIRT) program, the Army's research initiative focused on partnerships with HBCUs and minority institutions		

Table 4.1—Continued

Service	Name	Level of Student/ Employee	Overview	Minority Participation	Number of Participants	Description	Assessment	Annual Cost
Army	In-House Research	Researchers	Army research			Much Army research takes place within military institutions, under Army Materiel Command/Research, Development, and Engineering Command; Engineer Research and Development Center; U.S. Army Medical Research and Materiel Command; and the Army Research Institute		
DoD	Science, Mathematics & Research for Transformation Defense Scholarship for Service (SMART)	Undergrad and graduate school	Scholarship-for-service program	~20% (non-caucasian)	~150 new students selected annually (Since FY2005, ~1,130 students have participated in SMART.)	SMART is a scholarship-for-service program that provides support to high-performing U.S. graduate and undergraduate students in 19 academic STEM disciplines identified by DoD as areas of future workforce need.	82% of the students who have completed their service commitment are still employed by DoD beyond their original service commitment.	$25,000–$38,000 annual stipend; full tuition; book allowance; and health insurance reimbursement.
Army	Engineer and Scientist Program (ESEP)							

Table 4.1—Continued

Service	Name	Level of Student/ Employee	Overview	Minority Participation	Number of Participants	Description	Assessment	Annual Cost
Army	In-hours Laboratory Independent Research (ILIR)							
Army	Presidential Early Career Award for S&E							
Army	National Science Center (NSC)							
Army	West Point Center for STEM Education							

Table 4.1—Continued

Service	Name	Level of Student/ Employee	Overview	Minority Participation	Number of Participants	Description	Assessment	Annual Cost
Army	AEOP—Future Workforce Initiatives	K–12	Develop the future workforce	Minorities are one of the target populations		Includes a variety of programs	Conducted by the Virginia Tech Education Assessment Division Comprehensive analysis using quantitative and qualitative data • Application tool • Pre/post-surveys • Focus groups that target students, teachers, administrators, and parents • Provides the program administrators feedback on whether or not they are meeting their program objectives Provides the funding office information on effective/non-effective program impact on the Army's STEM education goals and objectives	

Table 4.1—Continued

Service	Name	Level of Student/ Employee	Overview	Minority Participation	Number of Participants	Description	Assessment	Annual Cost
Army	Current Workforce Initiatives	Ages 17+	Develop and retain the current workforce			Increase value of human capital through STEM competency Foster an agile workforce Retain highly competent talent		
Army	Gains in Education of Math & Science	K–12	Hands-on experiences			Part of the Army Educational Outreach Program—Future Workforce Initiatives		
Army	Mobile Discovery Center	K–12	Hands-on experiences			Part of the Army Educational Outreach Program—Future Workforce Initiatives		
Army	Junior Solar Sprint	K–12	Hands-on experiences			Part of the Army Educational Outreach Program—Future Workforce Initiatives		
Army	eCYBERMISSION	K–12	Competitions			Part of the Army Educational Outreach Program—Future Workforce Initiatives		
Army	Junior Science & Humanities Symposium	K–12	Paid internships			Part of the Army Educational Outreach Program—Future Workforce Initiatives		

t STEM Initiatives   55

Table 4.1—Continued

Service	Name	Level of Student/ Employee	Overview	Minority Participation	Number of Participants	Description	Assessment	Annual Cost
Army	Laboratory Apprenticeship		Paid internships			Part of the Army Educational Outreach Program—Future Workforce Initiatives		
Army	University Apprenticeship		Paid internships			Part of the Army Educational Outreach Program—Future Workforce Initiatives		
Army	SEAP College Qualified Leaders (CQL)		Focus near-term hires through research					
Army	University Research Apprentice Program (URAP)		Focus near-term hires through research					
Army	National Defense Science & Engineering Graduate Fellowship (NDSEG)	Graduate school	Focus near-term hires through research					

Table 4.1—Continued

Service	Name	Level of Student/ Employee	Overview	Minority Participation	Number of Participants	Description	Assessment	Annual Cost
Army	Multidisciplinary University Research Initiative (MURI) and University-based Centers		Focus near-term hires through research					
Army	Presidential Early Career Award (PECASE)	Employees	Focus near-term hires through research					
Army	West Point Cadet Program		Growing existing workforce capabilities					
Army	Section 219 funded efforts		Growing existing workforce capabilities					
Army	Army Civilian Training, Education and Development System (ACTED)-CP 16	Employees	Growing existing workforce capabilities					

Table 4.1—Continued

Service	Name	Level of Student/ Employee	Overview	Minority Participation	Number of Participants	Description	Assessment	Annual Cost
Army	Section 852 Funds for Acquisition Training	Employees	Growing existing workforce capabilities					
Army	In-House Laboratory Innovative Research (ILIR)	Employees	Growing existing workforce capabilities					
Army	Engineering and Scientist Exchange Program (ESEP)	Employees	Growing existing workforce capabilities					

CHAPTER FIVE

Conclusions and Recommendations

The 2012 DoD STEM Diversity Summit was coordinated by ASD(R&E) and ODMEO to support recent national and DoD policies related to federal workforce diversity. Our attendance at the Summit allowed us to develop several key questions and answers that may assist DoD in achieving its goal to increase the diversity of its STEM workforce. As noted earlier, this report does not evaluate DoD diversity and inclusion goals, but instead offers guidance on how to achieve those goals based on examining relevant literature and data, conference sessions, and conversations with senior DoD leaders.

Because of the state of research in this field, recommendations are inherently somewhat tentative: Most organizational change literature is highly theoretical, and, to the extent that there is empirical work, it tends to show that certain types of policies are statistically associated with certain results rather than causing those results. Despite this, our review of current STEM- and diversity-related national and DoD policy suggests there already exist substantial grounds on which to base further action. Overall, DoD STEM programs and policies are uncoordinated, their goals are not always clear, and they lack the kind of rigorous metrics needed to assess outcomes and on which to base improvements.

These findings are not based on complete analysis; we offer only a first response to Summit proceedings. However, given the limitations of this work, we can recommend *first and foremost that DoD articulate which aspects of diversity it will prioritize and set specific goals*. Programming, outreach, leveraging differences to enhance performance, and hiring practices are just some of the areas to better address once specific goals have been set. Significant organizational change research stresses the important of specific goals (Fernandez and Rainey, 2006).

The initial findings, especially those related to program assessment, suggest a second recommendation: that DoD work toward coordinating efforts across the organization to reach the goal of increasing the diversity of its STEM workforce. By *coordination*, we refer to the synchronization of organizational efforts, including the efforts of DoD as well as supporting agencies and external stakeholders. Best management practices suggest that building partner capacity and cooperation across a large organization and with partners can increase that organization's effectiveness in reaching goals, such as recruiting and managing a diverse workforce (Joint Chiefs of Staff, 2006; Mashaw, 2006; Alexander, 1995). Further, coordination among partners' efforts may improve efficiency and reduce costs through program sharing and lessening overlap.

As an employer, DoD has been recognized as exceptional in its enormity; established recently in the press as the largest employer in the world (BBC News, 2012), DoD consists of approximately 3.2 million personnel, including active duty service members, guardsmen, reservists, and civilian support. DoD is in a position to model how diversity is increased and managed in a large organization. Best management practices suggest that leaders promoting coordination across large organizations should do so in incremental stages, as building relationships, assessing efforts, and consolidating resources take time (RAND National Defense Research Institute et al., 2010).

Based on best practices, we offer the following managed-change plan to bring together the various efforts designed to promote a diverse workforce, including a STEM workforce. We recommend that DoD take action incrementally and in order: in the short term (1–12 months), mid term (1–3 years), and long term (4+ years). Table 5.1 depicts the timeline and intended participant (internal or external) for each of the recommendations.

DoD should act according to this incremental internal-external approach for two reasons. First, as we described earlier, change-management literature suggests that a few new policies or programs

Table 5.1
Recommendations Can Be Grouped Along Two Dimensions

Timeline	Internal Alignment within DoD		External Alignment with Partners	
	Within STEM	**Between STEM and Diversity**	**Within Defense STEM Ecosystem**	**National Alliance**
Short term (1–12 months)	Rec. 1	Recs. 2 and 3	Rec. 8	Rec. 9
Mid term (3–4 years)			Recs. 4, 6, and 10	Recs. 5 and 11
Long term (5 years and beyond)		Rec. 12	Rec. 7	Recs. 11 and 13

released over the next few years will not generate sufficient momentum and resources needed to bring about the deep, long-lasting changes needed to reach DoD diversity goals. Successful transformations of organizations are rare, and they require sustained efforts by a guiding coalition of leaders to change organizational culture, followed by the ongoing collection of metrics accountability practices tied to those metrics (Kotter, 1996; MLDC, 2010).

Second, the challenges associated with increasing the diversity of DoD's STEM workforce are tied not only to the services and defense agencies. STEM skills are developed within the education community and are supported by other federal agencies, as Summit presentations by Historically Black Colleges and Universities and NASA demonstrated. Partners must be considered, as implementing changes that affect organizational systems as well as subsystems over the long term can result in closer alignment with a desired end state (Hannan, Polos, and Carroll, 2003).

Align Policies and Practices Within DoD

Short Term
Recommendation 1: Articulate DoD STEM and diversity goals and align policies and practices within the DoD STEM community toward achievement of these goals.

The definitions of diversity throughout federal and DoD policy documents differ from one another, and in all cases are much too broad to use in evaluating current programs or proposing new ones. Additionally, as we have shown in Chapter Four, OSD and the services have many programs with varying objectives. Many also lack clear assessment of returns on investment. As we present in Table 5.2, according to the National Academies, ASD(R&E) investment in STEM totaled $165.52 million in fiscal year 2011. In this new era of fiscal responsibility, it is essential that members of the DoD STEM community come together under a set of clear and specific strategic objectives regarding improving the diversity of STEM workforce. As we reviewed in Chapter Two, the recently completed DoD STEM strategic plan provides an outline for strategic goals, but one that lacks specificity. Organizational change literature recommends having clear and specific goals (Fernandez and Rainey, 2006). This is helpful in terms of both designing policies and later evaluating performance.

We believe that as DoD implements its STEM strategic plan and further articulates specific diversity goals, it should also collect information about its current activities. With that in mind, we developed a template for collecting information, based in part on the information provided to us by the Navy on many of its programs. The template enhances the ability to collect information necessary to program assessment: (1) the name of the program, (2) its target participants, (3) an overview of the program, (4) the number and percentage minority population (ideally broken down by racial/ethnic group), (5) a fuller description of what the program does, (6) any assessment information available, (7) the program goal, and (8) how much the program costs the organization.

With this information, DoD can summarize where its current resources are going and begin to think about how to assess current

Table 5.2
ASD(R&E) Investments in STEM

STEM Programs	FY11 Presidential Budget Request / FY11 Enacted ($ millions)	Targeted Group
National Defense Education Program (NDEP) K–12 Informal Education	18/11.2	K–12
Awards to Stimulate and Support Undergraduate Research Experiences (ASSURE)	4.5/4.5	Undergraduates
Science, Mathematics and Research for Transformation (SMART) Program	56.0/48.8	Undergraduates
Historically Black Colleges and Universities/Minority Institutions Program	15/17.3	Faculty, staff, and students of minority institutions
National Defense Science and Engineering Graduate (NDSEG) Fellowship Program	38.3/38.3	PhD students at/near the beginning of their graduate study
National Security Science and Engineering Faculty Fellowship (NSSEFF)	36.12/30.72	University faculty, staff scientists, and engineers of accredited, U.S. doctoral degree-granting academic institutions
Presidential Early Career Awards for Scientists and Engineers (PECASE)	Army: 5.1 Navy: 5.1 Air Force: 4.5 Total: 14.7 (enacted)	Outstanding scientists and engineers beginning their independent careers

SOURCE: National Academies (2012).

programs and whether current areas of involvement are consistent with its goals.

We also recommend, as a next step, that DoD provide a similar template for program assessment. In order to evaluate the current set of programs, it must be possible to compare them. As it stands, such comparison is not possible because some programs may not be currently conducting assessment, and other programs, despite having similar goals, are assessing different factors. Devising a standard for assessment, such as the example shown in Table 5.3, should go hand-in-hand

Table 5.3
Standardized Template to Collect Program Information

Name	Participants	Overview	Service	Minority Participation	Number of Participants	Description	Assessment	Goal	Costs
SeaPerch	Middle school	Middle school robotics competition	Navy	45%	Number of students per year?	"Gateway" robotics program for middle and high school featuring a submersible remotely operated vehicle (ROV). Through robotics and underwater ROVs, students learn about career in naval architecture, and marine, ocean, and naval engineering.	Participating in SeaPerch increased interest in studying engineering in 25% of middle school and 30% of high school students. Program improved problem solving confidence in 34% of middle school and 43% of high school students.	?	Annual costs: $1,400,000 from ONR, $750,000 from SYSCOMS and NDEP

with determining DoD's goals for these programs. For instance, if the decision is made to narrowly focus on recruiting, as recommended by the National Academies' 2012 study, assessment mechanisms should focus on the degree to which programs change their participants' likelihood of later working for DoD as a STEM professional. If the decision is made to focus on increasing minority representation among those holding degrees in STEM fields, assessment mechanisms should focus on the degree to which programs change their participants' likelihood of earning a degree in STEM fields. By using standardized metrics, it will be possible for DoD to compare current programs to determine where to focus its resources.

Recommendation 2: Establish closer working relationships between ASD(R&E) and ODMEO in the short term and between AT&L and P&R in the long term.

By co-sponsoring the DoD Diversity STEM Summit, ASD(R&E) and ODMEO recognize they must work closely together to improve the diversity of the DoD STEM workforce. ASD(R&E) understands DoD STEM needs and manages its investment; ODMEO spearheads DoD diversity efforts. Both organizations recently finalized and published DoD strategic plans in their policy domains, as we reported in Chapter Two. Currently, both organizations are developing their implementation plans, which is the opportune time for both organizations to closely align their plans for maximal impact of their efforts. Having multiple senior leaders show support for new policies and work together to shape implementation signals organizational commitment and increases the chances of successful organizational change (RAND National Defense Research Institute, 2010).

As ASD(R&E) and ODMEO establish a closer working relationship, they should engage their senior leaders (AT&L and P&R) for support. After reviewing best practices in private-sector firms and scientific literature on diversity management, the MLDC concluded that without top leaders' support there is very little chance any effort to increase diversity will succeed.

Recommendation 3: Focus on building a pipeline to DoD employment.

As ASD(R&E) and ODMEO align their implementation plans, we recommend that they also start to collaborate on their activities to focus on using them as a pipeline to DoD employment. Both organizations currently sponsor outreach activities and internships. Hence, they and their service partners may consider co-sponsoring outreach activities and coordinating internships for better return on investment. Additionally, DoD should develop closer links from outreach to internships, from scholarship to hiring. For example, at its outreach events to minority communities, DoD may consider more aggressively marketing the SMART program.

To complement these efforts, DoD should consider conducting objective and high-quality research and analysis to guide its STEM and diversity outreach efforts. For example, Figure 5.1 shows four-year universities and colleges with varying number of minority students majoring in STEM. The size of the symbol shows the number of minority STEM students in each college. In addition, we show universities and colleges with top 100 STEM graduate programs using the *U.S. News and World Report* rankings.

DoD might employ such spatial analytics to guide its allocation of resources across the country. One can expand the spatial analysis to include DoD science labs and locations with sizable DoD STEM professionals to leverage local resources in the outreach efforts.

There is a body of literature on optimal recruiting resource allocation for enlisted personnel that can inform this analysis. Some of this work focuses on resources like recruiting bonuses and how they affect enlistment and re-enlistment (Asch et al., 2010). The literature that is more directly relevant to this problem, though, is the literature focusing on placement of recruiting stations. This treats recruiting station placement as an optimization problem, taking as variables various factors such as the population age 17–21, unemployment rate, number of high schools, current number of military recruiters, and cost of operating a recruiting station in that area (Mehay, Gue, and Hogan, 2000; Martin, 1999). The question of where to place resources to recruit more minority STEM graduates is a somewhat different one, since it is likely

Figure 5.1
Minority Students Majoring in STEM Attending Four-Year Colleges and Universities with Top Graduate Programs in STEM

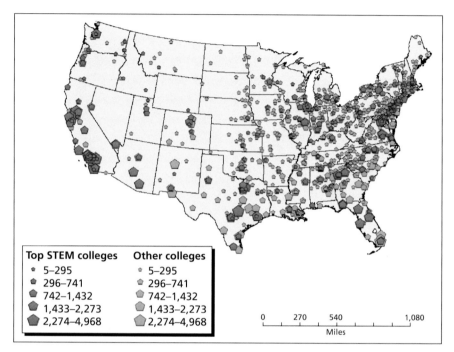

SOURCE: Based on data from the Integrated Postsecondary Education System (IPEDS) and *U.S. News & World Report*'s rankings of graduate programs.
RAND RR329-5.1

that these graduates are less tied to their local labor markets than the high school students and recent graduates that DoD targets for enlistment. Also, recruiting resources are less discrete than in the case of enlistment, where a major problem DoD must solve is whether to build or shut down recruiting stations. Resources for recruiting college students and graduates are likely more continuous and involve fewer capital investments: for instance, whether to send recruiters to a school's job fair. However, the same general type of optimization problem would be useful in guiding where to put these resources, and where minority STEM college students and graduates are located ought to be one of its inputs.

Mid Term

Recommendation 4: Expand strategic initiatives to include the Total Force.

As DoD implements its new STEM strategic plan, it should consider expanding the scope of some of its programs to include all members of the Total Force—including not only active and reserve components, but also civil servants and contractors. The briefings presented at the Summit concentrated solely on civilian programs. In fact, the National Academies noted the excellent investment that services make to develop their military members, recommending that DoD "upgrade education and training for the DoD civilian STEM workforce" by providing "similar opportunities afforded career military personnel and tailored to the needs of the civilian workforce" (National Academies, 2012, p. 11). Identifying programs from one component that could be successfully implemented in other components may help DoD to realize its STEM strategic goal.

Recommendation 5: Engage the Military Personnel Policy (MPP) and Civilian Personnel Policy (CPP) offices to overhaul recruiting of STEM professionals for all components.

The National Academies reported in 2012 that DoD is not the employer of choice among STEM professionals (National Academies, 2012). Talented STEM-capable individuals are in demand and frequently consider private-sector firms and academia as more attractive career options. The National Academies cite many potential reasons for this lack of interest, including but not limited to pay and speed of hiring decisions. AT&L and P&R should consider whether transforming aspects of DoD personnel policies could help it to achieve its STEM goals. ASD(R&E) and ODMEO may serve as catalysts for this organizational transformation, which will prepare DoD to meet emerging STEM needs and become an employer of choice of top STEM talents. AT&L and P&R should also coordinate with other offices within DoD that affect personnel policy, for both civilian and military components of DoD, to make sure their goals and strategies are aligned.

Recommendation 6: Establish specific goals for the representation of minorities and women in the STEM applicant pool.

A common maxim among evaluators is "what gets measured gets done." The importance of setting clear and specific goals so that the policy implemented corresponds with what was intended and implementing leaders can be held accountable is a theme within organizational change literature (Fernandez and Rainey, 2006). As we suggested in Recommendation 1, DoD should collect information about STEM programs using a standardized template, which include measures of participation by minorities and women in these programs. Because of legal limitations, setting goals for the number of hires or employees is not feasible. However, one potential option is to set goals for the applicant pool for DoD jobs and programs. As a part of its work, the MLDC carefully reviewed legal precedents for using metrics to guide diversity efforts. The MLDC concluded that it is legal to "develop goals for qualified minority [and women] applicants" (MLDC, 2011, p. 59). The MLDC also states:

> The Services have long employed incentive programs for recruiters to ensure that designated accession goals are met. This includes setting goals for the total number of accessions and goals for recruiting individuals with specific attributes, such as a high aptitude level (e.g., a high AFQT score) or specific skills or degrees (Oken and Asch, 1997). One way to ensure that there is a demographically diverse candidate pool from which to select applicants into pre-commissioning officer programs are to develop goals for qualified minority [and women] applicants. This strategy is currently employed by the Navy and the Marine Corps. . . . The goals would not be used during the actual admissions [or selection] decision but would help ensure that there is a demographically diverse pool from which to select new students [or participants] each year. (MLDC, 2011, p. 59)

An example of a goal might be that the applicant pool for specific jobs mirror the demographics of its occupation in the overall labor market, perhaps based on the sort of census analysis we perform in this report. It is possible that those standards are already being met in terms of racial/ethnic diversity, since the racial/ethnic composition of STEM workers within DoD already mirrors fairly closely the racial/ethnic composition of citizen STEM workers in the overall labor market. Another potential goal might be that the applicant pool for specific jobs mirror some weighted average of the demographics of the overall working-age population and the relevant labor market. We are not recommending a particular goal, but rather suggesting that DoD develop some type of measurable outcome, which will allow internal or external stakeholders to evaluate progress toward the articulated goals.

Long Term
Recommendation 7: Establish formal ties between policies and practices of AT&L and P&R (ASD[R&E]) and ODMEO).
Improving the diversity of the DoD STEM workforce cannot be done overnight. To see measurable changes, DoD must institutionalize closer coordination and collaboration between ASD(R&E) and ODMEO, as well as create innovative policies and practices that streamline recruiting and hiring policies and practices between AT&L and P&R. Consideration of impact on diversity should be a natural part of doing business at DoD.

Going Beyond DoD

Short Term
Recommendation 8: Consider establishing a Defense Diversity and STEM Advisory Council, representing the defense STEM "ecosystem," with an expanded mandate to provide oversight and advise the Secretary of Defense.
In the course of the Summit, Reginald Brothers recognized that "the need for STEM capabilities is growing" and "the [demographic] diversity in STEM is not high . . . this has become a national security concern." Brothers also offered a solution. A STEM "ecosystem," he suggested, could "leverage . . . creative minds wherever they are." Collaboration between DoD and "the entire system supporting it, including the labs and industry" could produce opportunity and enhance

diversity and inclusion, and as such yield "hybrid forms of new innovation" to ensure the nation's competitiveness.

Ellen Montgomery, program manager of the Air Force STEM Outreach Coordination Office, underscored the need to bring DoD's various diversity and STEM developmental efforts together through the concept of an ecosystem: "If we are not aligned and working this thing together here in DoD, we can't count on ourselves. We have a potpourri of functions."

We recommend that DoD consider establishing a Defense Diversity and STEM Advisory Council of members who have expertise in areas such as recruitment, diversity management, and STEM training. The council can give recommendations and feedback as the organization works towards fulfilling its workforce goal.

Recommendation 9: Be an agent for a national campaign.

In Chapter Two, we reported that executive and legislative national leaders have identified strengthening the quantity and quality of the national STEM workforce, as well as improving the diversity of the STEM workforce, by engaging underrepresented groups as a national priority. We suggest the DoD be a catalyst to reach its own goal by engaging with external stakeholders in a national campaign to improve the diversity of the STEM workforce. DoD is well positioned to take the lead in this endeavor because of its high visibility and the large size of its workforce, as well as the centrality of STEM to its overall mission. Additionally, it has significant interactions with other STEM employers and educators both within the government and the private sector, and DoD already co-sponsors STEM-related programs with some of them. Further, it has been established that the U.S. military is one of the most admired and trusted institutions in the country (Agiesta and Kellman, 2011). As the defender of the nation, DoD is in a unique position to speak about the importance of STEM. Additionally, DoD spends a considerable amount on advertising annually. For example, according to DoD Comptroller reports, the department spent $631.1 million on advertising for Total Force recruiting in 2012 (Office of the Under Secretary of Defense [Comptroller]/Chief Financial Officer, 2012). Given DoD's unique position in the nation and its level of advertising resources, the organization may be able to influence a national campaign. Before engaging in a full campaign, however, analysis on the cost-effectiveness of support and participation should be conducted.

Mid Term

Recommendation 10: Work with industry and academia to increase diversity within STEM professions.

The National Academies (2012) points out the importance of the industrial suppliers of DoD to meet the STEM challenges. The latest report states, "The testimony received by the committee and all of the data collected indicate that the major industrial suppliers of DoD are doing a good job of anticipating traditional and nontraditional STEM needs and acting aggressively to ensure that they have talent available" (National Academies, 2012, p. 117). In fact, the National Academies (2012) highlights several innovative initiatives by major defense contractors to create an "agile and adaptable" STEM workforce in their report. For example, the National Academies (2012) highlighted Lockheed Martin's Skunk Works program as one to emulate. In this vein, Lockheed Martin is also proving to be an exemplar organization in terms of its dedication to improving diversity of its STEM-oriented workforce. As Kimberly Admire, Vice President, Diversity, Inclusion, and Equal Opportunity, told the audience at the DoD Diversity STEM Summit:

> The private sector has an average of 5% of one's bonus as a result of diversity achievement. At Lockheed Martin, we recognize that what gets measured gets done, so we do have a comprehensive approach to measure what we do. That means not just recruiting externally, but what we do internally so that people are advancing into roles of responsibility that look like me. We have an employee survey that gives us an organization health index. We measure four things including diversity and inclusions. We then score those. We look at those by VP, business, and at a corporate level. We then require managers to create action plans based on these, and evaluate them in terms of how they do. And we tie this to compensation.

We suggest that DoD consider working together with industry and academia to increase diversity within STEM professions overall in order to fulfill its own workforce goal.

Recommendation 11: Support and track success of a national campaign to improve the diversity of the STEM workforce.

In the medium term, DoD should continue taking a role in promoting STEM careers and education with other partners to reach its STEM diversity workforce goals. As demonstrated by the Office of National Drug Control Policy (2010), best practices for promoting change include testing and evaluating message effectiveness. One possible role DoD could take in this regard would be to use recruiting campaigns to emphasize technical military occupations, the STEM skills that service members learn in them, and the usefulness of these STEM skills in nonmilitary occupations. Measuring and tracking the success of STEM-diversity programs like these will also be important for evaluating the performance of those programs and revamping them as needed.

Long Term

Recommendation 12: Enable the Defense Diversity and STEM Advisory Council to monitor policies and practices to increase the diversity of DoD's STEM workforce.

The effort to improve the diversity of the DoD STEM workforce must endure over the long term, and thus the Defense Diversity and STEM Advisory Council must be empowered to monitor and assess DoD efforts and advise the Secretary of Defense on how to improve STEM policies and practices. The Defense Advisory Committee on Women in the Services (DACOWITS) provides one model of the endurance needed to promote change over the long haul (MLDC, 2011). Other advisory council models should also be considered.

Recommendation 13: Sustain efforts to improve the diversity of the STEM workforce.

In the long run, DoD may be able to meet its STEM-diversity goals by sustaining efforts to increase diversity within its own STEM workforce, as well as continuing to contribute to national efforts to improve the diversity of the overall STEM workforce.

Summit Agenda

DoD Diversity STEM Summit
KPMG-1801 K Street, NW, Washington, DC
November 1–2, 2012
"A Business Case for Diversity and STEM"

AGENDA, November 1st

8:00–8:30 a.m.	**Check-In/Continental Breakfast**	
8:30–8:40 a.m.	Agenda Overview and Expectations	Dr. Lynn Scott, Facilitator
8:40–8:45 a.m.	Open Remarks & Summit Goals	Dr. Reginald Brothers, Deputy Assistant Secretary of Defense for Research
8:45–9:00 a.m.	Implementing Diversity and STEM Goals	Mr. Clarence Johnson, Director, ODMEO
9:00–9:25 a.m.	DoD STEM Strategy: Next Steps	Dr. Laura Adolfie, Director, OSD STEM Development Office
9:25–9:30 a.m.	Introduction of Dr. Sidney Ribeau	Ms. Karen Harper

9:30–10:15 a.m.	Education & STEM: Who Fills the Jobs	Dr. Sidney Ribeau, President, Howard University
10:15–10:25 a.m.	**AM Break**	
10:30–11:30 a.m.	Technology Driving the Future of STEM STEM/Technology Panel	

Dr. Karyn Trader-Leigh, Moderator
Ms. Camsie McAdams, Senior Advisor, STEM, U.S. Department of Education
Ms. Lucy Sanders CEO, National Center for Women & Information Technology
Dr. Cindy Moss, Director of Global STEM Initiatives, Discovery
Ms. Kenna Ose, VP Pearson Learning Solutions
Ms. Brenda Walker, Principal, Federal Advisory Services, KPMG

11:30 a.m–12:45 p.m.   Diversity & STEM: How to Get Best ROI
Panel of Chief Diversity Officers

Ms. Edie Fraser, Moderator
Ms. Kimberly Admire, Vice President, Diversity, Inclusion, and Equal Opportunity Programs, Lockheed Martin
Ms. Shari Slate, Chief Inclusion and Collaboration Strategist, CISCO
Ms. Neddy Perez, Chief Diversity & Inclusion Officer, Ingersoll Rand
Ms. Sherry Snipes, Chief Diversity Officer, American Institute of Architects
Ms. Susan Stith, Chief Diversity Officer, Express Scripts
Mr. Jose Jimenez, Chief Diversity Officer, CSC

12:45–1:45 p.m.	**Lunch**	
1:45–2:00 p.m.	Defining Our Investments in STEM	Dr. Laura Adolfie
2:00–2:30 p.m.	Department of the Army	Mr. Jeffrey Singleton, Director for Basic Research
2:30–3:00 p.m.	Department of the Navy	Dr. Anthony V. Junior, Naval BCU/MI Program Office
3:00–3:30 p.m.	Department of the Air Force	
3:30–4:00 p.m.	Summary/Open Discussion	Dr. Lynn Scott, Facilitator
4:00 p.m.	Adjourn	

AGENDA, November 2nd

8:00–8:25 a.m.	**Check-In/Continental Breakfast**	
8:25–8:30 a.m.	Welcome from KPMG	Ms. Candy Duncan, Managing Partner, KPMG
8:30–8:40 a.m.	Day 1 Summary	Dr. Lynn Scott, Facilitator
8:40–8:42 a.m.	Introduction of Dr. Johnson	Ms. Karen Harper
8:42–9:30 a.m.	Industry/Government/Education Partnerships: Best Practices	Dr. Roosevelt Johnson, Deputy Associate Administrator for Education, NASA

9:30–10:30 a.m.	DoD STEM Occupations: How Are We Meeting the Diversity and STEM Skill Requirements?	
	Dr. Karyn Trader-Leigh, Moderator Ms. Paige Hinkle-Bowles, Principal Deputy, Civilian Personnel Policy, Office of the Undersecretary of Defense, Personnel & Readiness Mr. Lou Cabrera, Comptroller/Director of Administration and Management Dr. Warren Lockette, Deputy Assistant Secretary of Defense for Clinical & Program Policy Dr. Valen Emery, Army Research Laboratory's Outreach Program Manager Vice Admiral Joseph D. Kernan, Military Deputy Commander, U.S. Southern Command Ms. Marilee Fitzgerald, Director of the Department of Defense Education Activity	
10:30–10:40 a.m.	**AM Break**	
10:40–11:00 a.m.	Leveraging DoD's STEM Education and Outreach Portfolio Future STEM Talent: SMART	Dr. Laura Stubbs, Director, S&T Initiatives
11:00–11:15 a.m.	STEM Outreach: K–12 STARBASE	Mr. Ernie Gonzales, OASD, Reserve Affairs
11:15 a.m–12:15 p.m.	Diversity and STEM: A University Approach Toward Achievements	Dr. William Harvey President, Hampton University (**invited**)
12:15–1:30 p.m.	Working Lunch/STEM & Diversity Integration	Open Discussion
1:30–1:45 p.m.	Wrap up/Adjourn	Dr. Lynn Scott, Facilitator

Summit Notes

Agenda Overview and Expectations, Dr. Lynn Scott

Throughout this conference, it's important to think about the following things: the business case for diversity and STEM, including how diversity can be positioned as a valuable component of a STEM strategy. You must make the dollars and cents case. Also, STEM education in schools, or how DoD can support the growth of K–12 students in STEM disciplines and educators certified to teach STEM curriculum, as well as the increase of postsecondary graduates in specialized STEM fields. Finally, think about this from the perspective of supporting a human capital pipeline—how can DoD support pathways to STEM occupations?

Opening Remarks, Dr. Reginald Brothers

I've been thinking a lot about innovations. There are three different models for innovation: the producing model of innovation, where Steve Jobs says, "I think everyone's going to like this iPhone." Second, the open source model. Third, the user/producer model, where people who are involved with the use of technology are in the creative process. I'm very interested in this model. As we think about STEM diversity, our goal is to increase our ability to innovate in this country. One example of user/producer model: a bunch of kids in California who wanted to ride their road bikes up and down mountains. They found that when they combined them with motorcycle components, it worked. This is now a big business.

Right now in STEM education, we have convergence in the fields of science and engineering. But we don't necessarily have curriculum that talks across these areas in a systematic way. DoD has several STEM priorities, including cyber warfare and electronic warfare. But those things are now converging. It's the same with quantum physics and information theory. When we think about diversity, we need to think of it across people, but also across disciplines. So how do we come up with set of actionable recommendations that can help us as a country incorporate diversity in a more powerful, comprehensive way?

Implementing Diversity and STEM Goals, Mr. Clarence Johnson

The DoD definition of diversity is the recommendation MLDC put forth. Diversity, in the STEM workforce as well as the overall workforce, is important. One third of DoD jobs are STEM-related. There's a demographic change. We're in a battle for talent. We need to get the best and brightest our nation has to offer.

In November, the President sent down an executive order to coordinate diversity in the federal workforce. We had a requirement to fulfill, and we worked with the services and submitted a diversity strategic inclusion plan. One of the ways we're working to achieve recruiting and retention is to increase the number of groups we engage with to make students more aware of careers in federal service. Again, the demographic makeup of the country is changing. To be inclusive, we need to take advantage of underserved communities. Young folks, even in elementary school, can learn to pursue STEM fields. Teenagers are often discouraged about their STEM skills or by peer pressure, particularly if they're minorities. I recently looked at a National Academy of Sciences study on diversity in the DoD STEM workforce. I noted recommendations for huge potential in the market for minorities and women. I am hoping this Summit provides insight in this area.

DoD STEM Strategy: Next Steps, Dr. Laura Adolfie, OSD STEM Development Office

The idea behind the STEM Executive Board was to corral all of the different moving pieces the department has to consider with STEM training and outreach. There are about 230 STEM programs. The criteria for inclusion in this group was a budget of at least $300,000, and it had to fit in certain bins—minority-serving institutions (MSIs), K–12, etc. I'm the director of the STEM development office, which supports the board and department in general and also runs the national defense education program. This includes the SMART program, scholarship for service, and K–12 program, which primarily is executed at our laboratories and facilities. We also have a basic research program, with researchers as well as students.

There's no one definition of STEM, even across the government. For instance, the NSTC [National Science and Technology Council] federal agency plan does not include social science, psychology, medicine, or information technology. It's very narrowly focused. Be careful looking at the numbers, and go to the source for their definition of STEM. For us, STEM includes military and civilian, occupations, and disciplines. Specifically, for the STEM Executive Board, we look at various occupations—computer science, life science, physical science, social science, engineering, and STEM technicians, as well as health practitioners, etc. A lot of our workers are former military, transitioning into civilian workforce.

Our three goals are to attract, develop, and retain a highly competent DoD STEM workforce based on DoD requirements, maximize the effectiveness of STEM investments, and codify DoD STEM policy.

Education & STEM: Who Fills the Jobs, Dr. Sidney Ribeau, Howard University

At Howard, there's a long tradition of graduating high achievers in STEM fields. We graduate the largest number of African American bachelor's degree holders who go on to earn STEM doctorates. Jobs in STEM fields will be growing, so we're growing our STEM pipeline. By 2018, 17 percent of jobs are expected to be in STEM fields, and we must ensure that entire pipeline is filled with students from a variety of different backgrounds. About three-quarters of STEM jobs are held by white workers, with 6 percent each for African American and Hispanic workers. We can do better. Women are also underrepresented. Although women hold close to half of the jobs, only 24 percent of them are STEM-related jobs. Although women make up half of college grads, they hold disproportionately fewer STEM degrees. At Howard, 66 percent of our students are female, about 12 percent major in STEM fields.

Howard has decided that STEM in education is critical: We've established a STEM council and made physical infrastructure a priority. We break ground on a new science building in 2013 that will bring together researchers from health sciences, engineering, and biotech. In addition to STEM, health care disparities are a major area of interest for us in our research. We have a new PhD program in computer science, and we have in the pipeline more African American PhD students in computer science than any other program in the country, at five. We have programs that send students to developing countries on STEM projects. For instance, students built equipment in Kenya for water purification and educated villagers. One of the greatest challenges in higher education isn't students' capabilities—it's connecting the dots to show them how to make a difference. I don't think Howard students are unique in this regard. These programs are designed to connect interest in STEM with real application. We also focus on professional development for faculty, as well as faculty mentoring both students and other faculty.

Howard runs a charter school, M2, which has been very successful, based on test scores and what high schools the students get into. The teachers are not traditionally trained. They have to get traditional certification, but we have teachers who were engineers, who were accountants, but wanted to go back and teach. These students come from across district, and there is no difference in male/female performance levels. We believe if we don't get to these students at a young age, they won't be prepared to be in the pipeline for Howard.

What I say to you in closing: There are a number of things we're doing on campus, but we could do more. Via partnerships with DoD and other federal agencies, we can come up with models, faculty members, and enthusiastic students. The greatest barrier isn't the qualifications of our young people—it's our inability to find ways on our own campuses to work together. We're not going to reach the goal unless we include everyone—women, Latinos, African Americans. Those groups have the capacity to do well in those fields; they just need an environment which nurtures their success.

Panel 1: Technology Driving the Future of STEM

Panelists spoke about ways that technology is changing how educators interact with students, including opening up learning materials to students all over the world. Examples include a system called Open Class provided by Pearson, which facilitates open online classes, as well as short video clips on real-world problems, followed by virtual labs where students play a game to solve a problem, which Moss brought up.

Stern brought up how the National Center for Women and Information Technology (NCWIT) is working with various groups to keep women and minorities in the STEM pipeline at all levels, including the workplace, with a series for helping supervisors address unconscious bias, as well as resources for evaluating mentoring programs. Moss talked about how Global STEM Initiatives provides math and science content and training to elementary school teachers, who may not have taken courses in those areas.

Teachers are widely held responsible for students' standardized test score results in math and reading, but science results are much less emphasized. Panelists discussed possible ways to remedy this, including content that could be added into curricula.

On the topic of how to increase number of African American and Latino STEM PhDs, panelists brought up the importance of mentoring and having a culture supportive of women and minorities, including by showing students that people who look and sound like them are in STEM fields. The conversation also came back to open online resources. For instance, MIT put STEM resources online, and found that about 80 percent of participants were male. However, among MIT's high school users, about half are female. Participants argue that having more women in STEM fields is an issue of national security and competitiveness.

Panel 2: Diversity & STEM: How to Get Best Return on Investment

Fraser brought up what a big change it is that diversity officers now report to CEOs and boards of directors.

Admire shared the Lockheed Martin perspective on diversity and STEM. This is a focus, first, because to have the innovative capability to provide the solutions Lockheed Martin's customers need, the company needs to have the hearts and minds of everyone in its workforce engaged, and it needs to be a diverse workforce. The second reason is that, to sustain as an organization, the company has to focus on diversity and inclusion because there's a shift in the demographics of our country. The Hispanic population is growing, and Lockheed Martin needs to have that pipeline.

Jimenez proposed that the big, multidisciplinary STEM problems, like big data and cloud computing, can't be solved without a diverse STEM workforce. In order to fill the vacant jobs requiring technological folks, African American, Hispanic, and female populations must be involved. It's not possible to separate their STEM requirements from diversity requirements.

Snipes brought up how 80 percent of registered architects are white males, and with the retiring of the baby boomers combined with shifting demographics, there will be a problem.

Baskerville spoke about how the National Association for Equal Opportunity (NAFEO) can help the government and private organizations attain diversity because their institutions are graduating a lot of African Americans, including 32 percent of the African Americans in sciences and engineering, and more than that at the PhD level. NAFEO also works with HACUs and the American Indian Higher

Education Consortium. She challenged the idea that if you have diversity, you don't have excellence, and argued for the need to start from the understanding that you can't have excellence without diversity.

Several panelists spoke about public-private partnerships. Sandia National Laboratories partners with industry, academia, nonprofits, and the Department of Energy. Programs include K–12 programs showing students STEM jobs. Jimenez argued for the need to measure ratios of conversion to DoD employment, and mentioned variance between conversion ratios of intern programs across the federal government.

Lockheed Martin has an IT program that works with youth, giving them jobs after they finish high school. One of the lessons the company learned is that it had underestimated the interpersonal and communications skills necessary for students to learn, like the importance of being dependable and reliable.

Stith noted the importance of early intervention and discussed Express Script's pharmacy program, which teaches 8th- and 9th-graders about the pharmacist career and, after graduation, accepts them into its pharmacy program.

Snipes discussed AIA's partnerships with other organizations to introduce kids to architecture.

Snipes talked about the need to help students make the connection between studying in STEM fields and the work they would be doing, as well as continue to encourage them after they get to college.

Jimenez discussed the benefit of having diversity metrics and incorporating them into employee evaluations, noting that this gets employees to make changes. Admire discussed how at Lockheed, an employment survey measures various organizational outcomes, including diversity and inclusion, with managers being required to create action plans based on the results, and then evaluated based on the outcomes of those. Compensation is tied to those outcomes.

Baskerville talked about the need to make a commitment to the diversity part of the interview process as well as the evaluation process, arguing that managers need to buy into the vision of a diverse DoD. She mentioned cuts in DoD funding to HBCUs, noting that the funding cuts and attempts to move programs out of DoD and into the

Army marginalize diversity and marginalize the relationship between DoD and the HBCUs.

Various panelists spoke about the need to make diversity seen as a business imperative or mission effectiveness issue, not just an HR function.

Defining Our Investments in STEM, Dr. Laura Adolfie, OSD STEM Development Office

The NSRC pointed out that DoD needs to strategically focus its STEM investments in the K–12 Department of Defense Education Activity (DoDEA) system. We focus much of our investments in the laboratory system.

DoD created five-minute videos for middle school students in which scientists and engineers talk about all of the innovative work they're doing. The National Science Teachers Association will be posting some of them for students.

There are about 90,000 students who are military-dependent children in DoDEA schools. Robotics is very popular with students. We also talk about the business of STEM communications and marketing. Investments are focused in learning, engagement, teachers, post-secondary, STEM careers, STEM system reform, institutional capacity, and research. We also engage our midshipmen and cadets in both the STEM experience and being mentors.

We're focusing now on data quality—assessment and evaluation. Both what DoD and senior leaders are interested in is outcomes and impact.

Regarding whether DoD STEM education funds have a diversity focus, in the National Defense Education Program, we do support Title I schools. That would be one example of how we serve underserved groups.

Department of the Army, Mr. Jeffrey Singleton

The mission of Army research is to design, develop, deliver, and sustain products for the soldier—and, really, the individual soldier in the small squad. The top challenges are greater force protection for our soldier—individually, ground vehicles, air vehicles, and bases, including forward operating bases (FOBs). Our soldiers are carrying way too much weight, and it's wearing out their bodies. We need research to bring down that weight. Other challenges include providing supplies to small FOBs, preventing and treating traumatic brain injury, and making soldiers more resilient.

The Army's definition of STEM includes social sciences and behavioral sciences.

We have various research initiatives with universities. These include university-affiliated research centers for research we don't have the capacity to do internally, like the Institute for Collaborative Biotechnology and the Institute for Creative Technology, where we partner with USC and the Hollywood gaming industry to do immersive training.

We also partner with the HBCUs and MSIs via our University Single Investigator Program. We work with Howard, Hampton, North Carolina A&T, and Delaware State. We want a broad space of ideas—it can't just be MIT or Stanford, because you get trained to think alike there. We want a wide set of ideas, so we invest in these other schools.

We also have in-house research, under five separate commands. We have about 11,000 scientists and engineers in our labs. Not all have degrees, many are skilled technicians.

We have funding, like for broadening the future STEM talent pool, as well as for research programs with the goal of hiring students afterward, including summer programs. We fund a lot of grad students. We also use Section 219 funds from Congress to focus on existing workforce capabilities, including long-term training like sending people back to school and lab improvements.

Recently, we started working with Virginia Tech on assessment tech. They help us with pre- and post-survey assessment, focus groups, etc., so that we can decide what's effective.

Department of the Navy, Dr. Anthony V. Junior

At the Naval STEM forum in June 2011, the Secretary of the Navy said he wanted to double investment in STEM. He also wanted to have programs that were portable, so they could be replicated across the country.

The National Academy of Sciences report said the low-hanging fruit in the STEM pipeline are minority students who have aptitude and capability but not opportunity. In the Navy, we focus on how to attract that population. We also want programs that are relevant to the Navy need.

Most of the funding is at the collegiate level—undergraduate through PhDs. We give scholarships and fellowships.

One of the things we track is diversity in our K–12 programs. If you have programs welcoming these folks in, you get them in the pipeline and then the question is how to keep them as they traverse through education and then come work for us. We bring in students to do research at our labs. We have this Naval Research Laboratory STEM program. We looked at participation—it's about 15 percent women and minorities. I think this is because our employees just weren't familiar with minority institutions. I worked with the CEO of the Naval Research Lab to set up a minority institutions research program to give them the same experience.

SeaPerch is a middle and high school program in which kids build remotely operated underwater vehicles. This year we partnered with Naval Recruiting Command and looked at how to give the military side of house greater participation in terms of serving as mentors and coaches. This program increased student interest in studying engineering as well as in confidence in their problem-solving skills. We also assess whether it improves student understanding of naval careers.

Iridescent is a New York and Los Angeles after-school experiential learning program. It has 10 Navy-relevant modules, as well as other activities, for 3rd- through 7th-graders. Half of the participants are female, and 80 percent of students after assessment want to pursue an education and career in STEM fields. We also have a science program with students in St. Louis.

We have a program that's a one-year master's in systems engineering program, with 12 students. We pay for a full scholarship. The curriculum is from the Naval Postgraduate School, so it's a Navy-relevant master's degree. When they finish, they have guaranteed employment with the Navy, and I think we only lost one student from the past three years. It's modeled after the SMART program.

We're working with universities to partner them with minority institutions. The idea is that an HBCU pairs with a research institution, works on research, then go after dollars together. For instance, Sonya Smith, chair of Electrical Engineering at Howard, works with Penn State with funding from Naval Sea Systems Command.

Department of the Air Force, Ms. Ellen Montgomery

We're still in our infancy in terms of our programs. The mission of our office is to be the focal point for all Air Force STEM activities—to create STEM outreach strategy and metrics, as well as to serve as the single clearinghouse focal point to keep efforts from being duplicated.

In Fiscal Year 2011, we had about 20 Air Force locations with active STEM offices. We reached 900 schools, 4,000 teachers, and 106,000 students.

We've done gap analyses, where we looked at where we have large STEM populations and large minority populations, and identified areas of interest. For instance, we now have programs in Los Angeles at Air Force bases.

We've been to 57 installations to talk about what events are occurring there. We've also talked to them about Air Force STEM policies, and shared best practices.

When we sought out proposals, we made them provide metrics for how they would measure success. We had about 120 programs that were approved, for $2.6 million in funding.

We also now have a website, with information and points of contact for our various programs.

Our office is also responsible for establishing outreach opportunities in the national capital region.

Through Air Force labs, we work with HBCUs. We need more females and minorities in STEM. We're trying to actively recruit in these large, diverse populations at all levels of education to get them involved in STEM. We are cognizant that we have underrepresented communities in STEM. We've seen about a 5 percent growth in our minority male, minority female, and nonminority female STEM workforce.

One of our initiatives is to target high school teachers from underrepresented communities. We just had 25–30 teachers come in to Howard to learn about Air Force initiatives. These materials camps are a joint effort between Air Force STEM and Air Force diversity offices. We run these camps out of various universities by partnering local Air Force bases with universities.

We are also running cyber summer camp, where the kids had to build computers. It was a large success. One of the lessons learned was—next year, we are focusing on computer safety for parents as well.

We're trying to track diversity data, but we're still figuring it out. We are trying to voluntarily track gender, race, socioeconomic background, physical abilities, and language abilities for our programs.

There's no direct relationship with Dr. Taylor, the Deputy Assistant Secretary of the Air Force for the Strategic Diversity Integration.

Q&A Discussion

There is a compelling reason to better integrate STEM and diversity in DoD, the labs, and the private-sector companies that work with DoD. In some private sector companies, diversity is code for compliance. I've seen too many people with hiring authority who still have a negative mentality about HBCU engineers.

In order to have better coordination, we need to have the organizational structure to facilitate coordination and leadership.

Scott brought up how it would be helpful to talk more broadly about the business case for diversity.

Brothers requested data on changing demographics, the growing need for STEM, and diversity in STEM fields. This may become a compelling business case. What does that look like—the need for STEM, how that's growing, with the growth in minorities and the growth in need for STEM? That trend becomes a national security concern.

Everything is changing. Let's think about manufacturing: When you put additive manufacturing with high-power computers with advanced data analytics, individuals can manufacture on their own. We have to educate young people on these things. What we need to get out of this session is not just recommendations, but recommendations that lead to policy suggestions.

For instance, if we need more mentorship, what policy changes do we need to think about? Are there things we can do policy wise to make it easier for students to do internships with labs? What can we think about for policies? That's the context of our discussion.

Our K–12 investments are diffuse. We're not clear where they go. That's not bad necessarily, but it depends what our goals are.

A new study on STEM concludes that it's difficult to predict what fields we'll need in the future. There's a current list of science and technology priorities, but will those be the priorities of the future? For the current priorities, maybe we invest in PhD programs, STEM, but if we're talking about a broad portfolio for the future, maybe we invest in K–12. This depends on our timeframe.

The projects we have now tend to be isolated, and not necessarily based in best practices.

Brothers concluded by saying, "I'd like the services to think closely about JROTC programs. I used to be a recruiter, and I remember visiting ROTC detachments. Particularly in urban areas, we have a lot of kids who are a captive audience. The data shows that about one-third join the military, but 98 of those are enlisting. It should be called JRETC! The report talked about DoDEA—that's a captive audience, too. But JROTC kids are, too. We're graduating 500,000 kids a year—why don't they have a STEM curriculum? It doesn't have to be a recruiting tool. They have a huge minority and women population."

Day 1 Summary, Dr. Lynn Scott

These issues should guide your thinking again as you listen to presentations today. First, the business case: We have to have an argument for why this is important. That means not just why this will save the world, but why the investment of our time and money will save the world. There are two sides to a business case. One is advocacy. But the other is whether you can do anything about it.

The second issue is the investment portfolio. Yesterday you saw the range of options in the portfolio. So where do you invest your time and power. K–12? University? With the services? A combination?

The third is return on investment. Are participants working for you, or are they leaving? This includes the whole ecosystem, including contractors.

DoD has a legacy of investment and science and technology since World War II that has continued ever since. How do you leverage that?

Some interesting issues from Panel One include the role of community colleges, as well as partnering with private industry, other departments, and philanthropists. The Gates Foundation is a great example. It has invested millions in education. There is also the need to assess existing programs for capacity and effectiveness.

The Army gave a presentation in which it was mentioned that the Army has 7,000 technicians and analysts. These are not just college graduates.

Industry/Government/Education Partnerships: Best Practices, Dr. Roosevelt Johnson, NASA

NASA wants to be a major player in STEM education. Currently, NASA has an interesting combination of programs. I'll give you some context for why we do certain things, and then examples of some of the implementation.

Our interest in the STEM workforce is that we need different people besides astronauts. About 65 percent of our workforce is STEM trained. It's in our interests to make sure that the STEM workforce is vibrant and vital.

NASA can do stuff no one else can do. We make big things look easy. Also, we can engage students—I have yet to see one who wasn't excited by space.

The internal NASA universe is complex and geographically spread out. That presents challenge for coordinating programs. Each component has educational programs with some autonomy, run separate from HQ. Coordinating everything is one of our challenges.

In the past, we had more money, so people could do things without as much coordination. Now, with reduced budgets and increased scrutiny, the impetus is that you have to be more coordinated.

How do we produce more undergrad STEM majors? As NASA we asked, how do we get a million new ones? Four out of ten 4th-graders aren't mastering math and science, so the challenge is not undergraduates—a bigger part of the challenge is going down to where there's a pool of students and building it up. There's a lot of heavy lifting to do if you start with this at the undergraduate level.

We try to inspire kids to do something in STEM. We want to get them excited. There is data showing students who are more engaged do better in STEM. They make different course choices and activity choices.

Our teaching core in this country isn't prepared for STEM education. In some places, we don't have STEM-trained teachers teaching STEM. This isn't an indictment against teachers.

We're trying to get to those kids who might otherwise not be reached. Some of our programs directly target diversity—but even in those that don't, we need to make sure we do make diversity part of that, while still being legal.

We can't do all things we'd like to ourselves, so we develop partnerships. We have partnerships with learners, educators, and institutions. Our 100K in 10 program is collaborative. We are working with all sorts of groups that do STEM teacher training.

There are unique things NASA can do. We had kids sending experiments to the international space station—elementary school kids. We also coach a lot of FIRST Lego competitions robotics teams.

Our engagement at the undergrad level is through Minority University Research and Education Program (MUREP) to underrepresented folks. We also have a space grant that engages universities, and The Experimental Program to Stimulate Competitive Research (EPSCoR) funding that's research-oriented. At least 60 percent of our education budget goes to undergrad institutions. The rest is K–12. And each center does very unique things in their region. There's a lot of activity around the Jet Propulsion Laboratory. The difficulty is keeping it coordinated.

There's significant value to diversity in decisionmaking context. There's research showing diverse groups of people make decisions differently and better. Industry has realized this. I believe it. You get different perspectives with diverse groups.

Panel 3, DoD STEM Occupations: "How Are We Meeting the Diversity and STEM Skill Requirements?"

Fitzgerald argued the DoD does a much better job of showing that mission failure results when you don't pull diverse groups of people together, and suggested a campaign centering around mission imperative. Also, when these young folks come into the workforce, they need to give them responsibility. They don't want to do filing. And private industry puts them on management teams immediately. So if we want to hang onto these folks, they have to come into the workforce and we have to give them responsible jobs and assignments.

Vallen mentioned the need to broaden the institutions they look at in terms of hiring. The DoD can't always hire from MIT, but has to go to a broader pool.

Lockette discussed the need for more communication between the science and technology community and the line community, and suggested the amicus briefs on the value of diversity for the military.

Brantley brought up how scouts try to find sports talent, but no one is scouting students into STEM fields. She mentioned that veterans leaving the military are a great resource to educate both the workforce that stays in, and K–12.

Fitzgerald Hinkle-Bowles and Brantley discussed efforts to work with veterans and bring them back into the DoD workforce and the workforce in general. Emery mentioned that the Army Research Lab has specific technical degree requirements, and so isn't always a good fit for service members, but there are opportunities for veterans in operations.

Fitzgerald brought up the figure that 1 in 5 children who graduate from high school are not qualified for the military. She argued for the need to start with pre-kindergarten. The United States is in the lower quartile in high school graduates, behind not just Japan and China, but also the Czech Republic. If children are not reading by third grade, their academic journey is compromised, as is their ability to enter the workforce.

Emery brought up the need to get very strong teachers in the college freshman science classes. Fitzgerald agreed about the need to inspire students, and brought up DoDEA courses in fields like gaming and green technology. DoDEA is also adopting common core standards.

Panelists also spoke about mentoring employees. Hinkle-Bowles mentioned development programs, and the expectation that senior leaders will be mentors. Lockette argued you don't need to incentivize mentors, but that a lot of people are interested in mentoring and either already doing it or need to be asked.

STEM Outreach: K–12 STARBASE, Mr. Ernie Gonzales, Reserve Affairs

One of the Office of the Secretary of Defense, Reserve Affairs, youth outreach programs is the STARBASE program. It works with 5th-graders in Title I schools around military facilities. The focus is on how to excite kids about science and technology in these places where we don't normally recruit.

The curriculum includes an introduction to computer engineering: Kids are engaged in technology, and need to learn math for their designs to work. We invested in 3D printers so they can design objects and have them printed, and show people what they've done.

We could make our program efficacy numbers better if we didn't go into Title I schools, but if you want to talk about diversity, you have to go into schools where it's difficult for kids to graduate.

There are 76 locations, with 70,000 students annually, with all components. The majority of the program is operated by the National Guard at a cost of $330,000, for people, equipment, and supplies. Installation folks are responsible for cost of facility—that's part of their contribution.

The STEM mentoring issue is how to keep working with these school systems to create a pipeline that's relevant to DoD. We created mentoring in these schools, and we limit mentoring to four students per mentor. We focus this around kids having a role or activity, like robotics or FIRST Lego. So we have former STARBASE kids participating in these competitions. But this may not connect with the DoD pipeline. These are not highly recruitable areas, so DoD isn't there. The goal is to reach 100,000 kids.

Diversity and STEM: A University Approach Towards Achievements, Dr. JoAnn Haysbert, Howard University

At Hampton, building our STEM programs has been a major focus. When we began to build our physics program, we initially began a doctoral degree program. We realized we need partnerships to do this. We began working with the Department of Energy, and also formed a partnership with NASA Langley where we hired two of their research scientists to work on the atmospheric and planetary sciences, which was an outgrowth of our physics department. We're the only HBCU with weather satellites. We're also working with Jefferson Labs, which is sponsoring faculty posts.

We're working with partners on a variety of other important programs, including breast cancer detection and our Hampton University Proton Therapy Institute. It's the largest one of its kind in the world, and it serves not just as a medical center but it conducts research, in addition to training our faculty as well as students.

We're also partnering with DoD on our Data Conversation Management Lab, which enables us to convert printed materials to electronic materials for the DoD.

One of the things we've always considered is ensuring we don't exclude women.

We want to be the top school in terms of graduating underrepresented minorities in STEM fields. We're now #3.

We're proposing a collaboration bringing together high schools. We think there ought to be summer programs for high school science, computer science, and math teachers. This will address teaching methods and new methodology.

Another element of the summer program would be a pre-freshmen program, for freshmen who plan to enroll in STEM. We'd also like to implement a program for rising seniors in math and computer science, which will focus on calculus. Another component is to have 9th- and 10th-graders focusing on career opportunities. The final summer program is for community college teachers. We hope to focus on exposing community college teachers to best practices about STEM.

The return on investment in our existing programs has been realized. Hampton is #3 in terms of educating black STEM PhDs, and #5 in engineering and science. Twenty-seven percent of our grads are in STEM, compared to 17 percent nationally. Our model has been successful. Next, we're going to actually become a research university, perhaps over the next 10 years.

Our STEM grads tend to either continue in advanced programs, or they take industry jobs. In terms of attrition, we're not losing a lot of students in the STEM areas. When we began, we had trouble keeping people in STEM areas. But we're better at it now. We conduct surveys. I make telephone calls to people thinking about changing majors.

Leveraging DoD's STEM Education and Outreach Portfolio, Dr. Laura Stubbs, OSD P&R

SMART is a scholarship for service program. Eligible candidates are U.S. citizens pursuing an associate's, undergrad, or graduate degree in a discipline critical to DoD, who maintain a 3.0 GPA and can get and hold a security clearance

There are 19 disciplines in STEM that qualify. There are two groups of participants—non-DoD employees, and DoD employees we're trying to retain.

It comes with full tuition and fees for up to five years in length depending on the degree. There's also a generous annual stipend: $25,000 for a B.S., and up to $38,000 for a Ph.D. And health benefits. They're also mentored.

We guarantee postgraduate employment. There's a service commitment of at least one year of scholarship to one year of service; three years of service for current DoD employees.

The 19 disciplines are based on what we hear from the services.

We'd like to grow the program, but in 2012, it was much smaller because of the budget.

Women make up about a quarter of participants, and white people about 80 percent. They're trying to get more women involved by engaging with affinity groups. We're also working on engaging with HBCUs and MSIs. We have folks from those colleges on our review panel, but final selections are made by the hiring organizations.

The program is very selective. In the last couple of years, there have been about 3,400 applications per year. In 2010–2011, about 300 were chosen. This year, only 134 were chosen.

Working Lunch/STEM & Diversity Integration, Open Discussion

Scott described the objective of this exercise as (1) determining how to talk about this issue, (2) determining the desired end state and how to get there, and (3) identifying two actions we'd like undertaken as a result of this that will be successful.

Scott continued that the business case can be framed in one of two ways—affirmatively in terms of gains, or in terms of threats. The second is easier: For instance, as a threat to the capabilities of the department. I see a threat in allowing this gap to remain or be increased between the line and the S&T community. This could lead to a reduction in support for the department, because the demographics of the country are changing in ways that, as power shifts, our political leadership will look at the department and ask, is it representative of the country as it should be?

The innovation threat, from China and India, is very important. We have to get demographics right to get the diversity. If we don't get results right, we'll have political risk and budget cuts.

In China, students understand the necessity of STEM for their nation. They know that to solve China's issues, they need STEM. We don't have that here. We need a national discourse that STEM is how we secure our nation and protect our quality of life. The DoD can lead in terms of messaging on this. Like, in the 60s, getting ourselves to the moon was a national narrative.

Scott mentioned some hypothetical pushback to that argument, stating that maybe the United States just needs a few genius entrepreneurs, not lots of people with skills in these disciplines. He also focused on the need to persuade people who don't share the same point of view on this.

If the supply of STEM workers increases, wages can fall, creating cost savings for the department.

We need to talk about getting the workforce we need. Everyone isn't going to be an honors student. So how do we focus on late bloomers? How do we provide the people we don't give those opportunities these experiences? For instance, the applicant pool for the SMART program is people who already have opportunities for support from elsewhere, not the underserved.

We already have existing relationships and pots of funds. One way we may use this is to leverage our relationships with minority institutions. For instance, we can fund students to get involved with DoD research, like via internships at the labs. So, now they can reach into communities for hiring that they're already familiar with because they do recruiting there. So you create this pipeline by using institutions. Through existing funding, we can ultimately create a pipeline for DoD labs.

I'd like to see our teachers feeling like they're part of the STEM workforce, and knowing how valuable they are to creating future scientists and engineers.

Regarding the business case for diversity, Change the Equation, which was established through this administration, now has 140 corporations signed on. Change the Equation refers to the lack of diversity in STEM fields. I'd recommend you get familiar with the institution. It's headed by Linda Rosen. They've gone a long way by identifying the business case.

Use the assets you have. The biggest asset department has in this area is its workforce. You saw this with Ernie's example of National Guard mentors. Getting them engaged with teachers and kids is a great strategy and it doesn't have that much to do with writing a check. Much of the assets we have to deploy are the humans already in the service.

Back to the demand side, we already have many STEM jobs that we can't fill with STEM workers. We have to start where the jobs are, where the career pathways are from the demand side. The HBCUs, Hispanic-serving institutions—they're all important—but if we partner with everyone we're talking about, the messaging is really clear if we can really excite students about careers.

Scott asked for responses to the argument that the DoD is already doing a lot of things, and it may not have the capability to enhance its collaboration with everyone, but needs to prioritize.

In the Army, there isn't segregation between the line and STEM. Leaders want soldiers who are smart enough from a STEM standpoint to utilize the technical equipment. That's an issue in all sorts of branches—ground forces, combat arms, support—they want STEM people. If you don't understand the communications equipment, you can't communicate. I was talking to a general officer who wants engineering graduates—not because of any equipment, but because he's seen how engineers problem-solve, their processes for that, and he wants that on the battlefield. It's the same reason why law schools and medical schools want engineers: the problem-solving process.

A note on solutions: The National Science Foundation (NSF) has a $7 billion budget to do what we're talking about. I've been there and asked why they don't have a strong partnership with the DoD, but I haven't gotten an answer. I'll ask it of DoD—why is there no strong partnership with NSF and their $7 billion to solve the challenges we've put on the table?

One business case that hasn't been brought to bear is about returning vets. Many with high school degrees or GEDs have been running submarine systems, but don't know where to get a job. They have well-developed human capital. This is a huge case for STEM diversity.

I'm working with API as a consultant. It's an association of gas and oil industries. Gas and oil aren't going away, as much as we might want them to, so they looked at gas and oil growth forecasts domestically. They then looked at job categories necessary for those growth areas. They're in the central part of United States. There are about six areas where they project growth in the billions of dollars. Then they looked at job categories necessary to sustain that growth—from professional to semi-skilled. And guess what—they're not all engineers. They're looking for people with high school degrees and certificates for $60,000, $70,000 jobs. Oh, and, API didn't even think twice about making the case that that labor pool is black and Latino. That was the assumption they made, that the workforce for that is black and Latino. So they did an analysis on the demographics. We need to get smart about that, about doing that match.

Scott asked, once you've convinced people with your business case that this needs to get done, how to invest funds.

Soldiers are coming out of the services with a technical background. When we learn engineering, normally the order is theory, application, employment. Service members have a different sequence—employment, then application. That's an excellent population if we have the money for getting them in school to learn the theory. How can we take these service members and get them into undergrad or grad programs?

Scott asked if there are opportunities for investment in new curricular models that would allow those service members to quickly make the transition through curriculum based on their needs, so they can get to level we need them to be at quickly. He suggested this investment could be directly through the DoD, or indirectly through the University of Pittsburgh and its distance-learning program.

Scott asked about how, once you have kids with all the skills you want, how do you get them to come work for you?

The Army Corps of Engineers is easily filling positions, but would like to improve the quality and expertise of the engineers it hires.

In DoD, there are a lot of retirement-eligible STEM civilians, so there's going to be a big gap. Hiring military members to come back may not fill that gap: We need to bring on civilians to fill existing and future jobs. In a bad economy, retention isn't an issue, but when it improves, it's hard to retain on both the officer and enlisted side for STEM skills. How do we reward people so they continue to stay and excel and be in leadership positions?

There aren't enough STEM minority leaders. There's a generational aspect—bench scientists may not want to become leaders, but leadership is critical. Women and minorities need leadership skills.

It's time to look at the current snapshot in terms of eligible retirees and reassess your requirements. People are staying now because of economics, but they're strategically waiting for their time to leave. Forecast what your individual needs are. Look at your forecast, look at your population. Yes, we need minorities, females, everyone. But do you really need a nuclear engineer? Do you really need an aero engineer? A mechanical engineer? Reassess your requirements and your needs. Look at who is going out the door and look at who is leaving. Also, one recommendation of the National Academy of Sciences report was that the department needs to do better job of developing the current STEM workforce. There's a new generation of kids coming on board who are not looking for the 25-year career. They're coming in for 3 years, then moving on. In addition to identifying gaps with the current workforce getting ready to leave, I submit that when we bring new kids on, we need to figure out how to retain them.

One of the challenges in the system is table of distribution and allowances (TDA) allocations. We have a set number of people, and we can't hire beyond that. How do you address the skill sets in your organizations? For example, in our laboratory we had no capability in cognitive and behavioral science. We grew that with SMART. That's a big challenge for us—TDA and how do you overcome some of those limits as technology changes. Another issue is culture. You need to put commitment for diversity as a performance metric, and have a tool to measure some of the things you need.

People like the National Governor's Association and Gates Foundation have been providing grants to do STEM on a state level. There are thousands of programs, but there's not enough communication between them. I think the DoD can be a big part of this. I'd encourage you to figure out what's going on—there are lots of people who want to partner with you.

We have to get our own house in order. I'm not saying partnering with other people isn't important, but I'm not sure if our house is in order. I don't know where we're spending our money. I've seen some opportunities in the past few days to get this sorted out. Let's just be more deliberate. I think if we want to partner with more than one HBCU or MSI, make them collaborate together. I'm more concerned about how we're collaborating internally, that we don't have a clear picture. There are opportunities from today, particularly with the strategic plan/vision, for us to collaborate so that we're not duplicating.

Scott asked to transition into the third area for discussion, suggestions for what we're going to do.

One of the things this department has failed to do is figure out what are we trying to solve. I think there are a number of different opinions. We're trying to get more kids in STEM degree programs. We're trying to get DoD workforce to look more like the U.S. population. We've got to get very young kids excited to increase the talent pool. But it's not real clear what we're trying to solve. We have to figure out real problem statement. What are we after here as an end state? We have to define that. Without that, we'll be doing things that are really disconnected.

Setting some time horizons with different goals might help us get at the endpoint.

Where are we going to invest our resources? Over the course of the past two days, there's been a lot of disconnect. Maybe we need to capture best practices, put them together—maybe on a website—so we can determine whether we can capitalize on those best practices.

There's not much knowledge about the thousands of disconnected activities. In the Navy, we spent a year and a half building a database of this. But there are a lot of legacy activities that someone has an emotional interest in.

There are obvious things that can be done to keep interns engaged and join the services. In some research programs, there's no service commitment, and a lot of them don't even know who writes the checks. I've made recommendations to take the programs without a connection to the service to make a tweak to make them more effective. There are a hundred low-hanging fruit opportunities we could do in the next 90 days.

Scott continued to suggest the need for a business case. Participants asked why this was necessary, arguing that the military, as well as the business and education sectors, are already convinced.

Scott replied that one assumption here is that DoD's problem is different from the nation's problem. But we're not insulated in being DoD from what's happening with rest of country. There may be some specificity with DoD investments. I think we need to consider how the DoD business case fits with the rest of the world.

The business case has been made. The issues for the nation are even more challenging for the military because we can only hire U.S. citizens. There's a lot of work to do on investment side.

As a military person, I'm ready to get on it with it. We've done enough research. We know what the challenges are. We spend a lot of time trying to solve world peace. There are a lot of opportunities for short-term work, what we can do in six months. What is that framework for our offices? We need to make the commitment to get back together. I think we have enough people here so that if we decide what we'll work on, we can do it. We've studied enough.

Articulating a business case would be a good move forward. We can certainly in six months put something forward. It won't be hard. Maybe put some synergy around what the services are doing with HBCUs and MSIs. Similarly, I think we can work with personnel folks, ATL, to take advantage of the current workforce. I think all of these things are useful. I think we can capture wins in all of them.

I agree about a business case to inform a message we can all stick to [concluded Scott]. This can also inform where we invest in. Let's talk about what I've learned this week: what I hear is that on this issue of capacity, pipeline—with students and teachers—there's been some debate about whether we have informed leadership, this line commander issue, from my own experience I'd argue there are some gaps. What I'm impressed with is this analogue to the National Football League. There are recruiters, scouts, Pop Warner. We don't have anything like that with STEM. So, there could be an NFL-like campaign not just to train and recruit players, but also coaches. But, in the short term, in order to get momentum, we need short-term wins. I think we can get that with messaging, and with the business case. Other things we can do now are pilots. What type? Mentoring? Something we can do that gets momentum.

References

Agiesta, J., and L. Kellman. 2011, September, 13. Poll: Americans trust military, but not Congress. Associated Press. As of June 10, 2013:
http://news.yahoo.com/poll-americans-trust-military-not-congress-165344167.html

Alexander, E. 1995. *How Organizations Act Together: Interorganizational Coordination in Theory and Practice.* Luxembourg: Gordon and Breach.

Asch, B. J., P. Heaton, J. Hosek, P. Martorell, C. Simon, and J. T. Warner. 2010. *Cash Incentives and Military Enlistment, Attrition, and Reenlistment.* Santa Monica, CA: RAND Corporation, MG-950-OSD. As of June 24, 2013:
http://www.rand.org/pubs/monographs/MG950

BBC News. 2012, March 20. Which is the world's biggest employer? March 20. As of June 10, 2013:
http://www.bbc.co.uk/news/magazine-17429786

Beede, D., T. Julian, D. Langdon, G. McKittrick, B. Khan, and M. Doms. 2011. *Women in STEM: A Gender Gap to Innovation.* Washington, DC: U.S. Department of Commerce Economics and Statistics Administration, #04-11, 2011.

Carnevale, A. P., N. Smith, and M. Melton. 2011. *STEM: Science, Technology, Engineering, Mathematics.* Washington, DC: Georgetown University Center on Education and the Workforce. As of June 10, 2013:
http://www9.georgetown.edu/grad/gppi/hpi/cew/pdfs/stem-complete.pdf

Cover, B., J. I. Jones, and A. Watson, 2011. Science, technology, engineering, and mathematics (STEM) occupations: A visual essay. *Monthly Labor Review Online,* 134(5): 12.

DoD—See U.S. Department of Defense.

Falk, J. 2012. Comparing the Compensation of Federal and Private-Sector Employees. Washington, DC: Congressional Budget Office.

Federal Chief Information Officers. 2011. *Opportunities Exist to Improve Role in Information Technology Management.* Washington, DC: U.S. Government Accountability Office. As of June 10, 2013:
http://www.gao.gov/new.items/d11634.pdf

Fernandez, S., and H. G. Rainey. 2006. Managing successful organizational change in the public sector. *Public Administration Review* 66(2): 168–176.

Galama, T., and J. Hosek. 2007. *Perspectives on U.S. Competitiveness in Science and Technology.* Santa Monica, CA: RAND Corporation, CF-235-OSD. As of June 10, 2013:
http://www.rand.org/pubs/conf_proceedings/CF235

Hannan, M. T., L. Polos, and G. R. Carroll. 2003. The fog of change: Opacity and asperity in organizations. *Administrative Science Quarterly* 48(3): 399–432.

Hough, L. M., and D. S. Ones. 2001. The structure, measurement, validity, and use of personality variables in industrial, work, and organizational psychology. In N. Anderson, D. S. Ones, H. K. Sinangil, and C. Viswesvaran, eds., *Handbook of Industrial, Work, and Organizational Psychology—*Vol. 1: *Personnel Psychology.* London: Sage.

Joint Chiefs of Staff. 2006, March 17. *Interagency, Intergovernmental Organization, and Nongovernmental Organization Coordination During Joint Operations.* Joint Publication 3-08.

Kotter, J. P. 1996. *Leading Change.* Harvard Business Press.

Langdon, D., G. McKittrick, D. Beede, B. Khan, and M. Doms. 2011. *STEM: Good Jobs Now and for the Future.* Washington, DC: U.S. Department of Commerce Economics and Statistics Administration #3-11.

Martin, Paul E. 1999. *A Multi-Service Location-Allocation Model for Military Recruiting.* Master's thesis. Monterey, CA: Naval Postgraduate School.

Mashaw, J. L. 2006. Accountability and institutional design: Some thoughts on the grammar of governance. In Michael Dowdle, ed., *Public Accountability: Designs, Dilemmas, and Experiences.* Cambridge University Press, pp. 115–156.

Mehay, S. L., K. R. Gue, and P. F. Hogan. 2000. *Recruiting Station Location Evaluation System (RSLES): A Summary Report.* Monterey, CA: Naval Postgraduate School.

Military Leadership Diversity Commission. 2010. *Change as a Process: What Business Management Can Tell Us About Instituting New Diversity Initiatives.* Arlington, VA. As of June 10, 2013:
http://diversity.defense.gov/Resources/Commission/docs/Issue%20Papers/Paper%2021%20-%20Business%20Management%20and%20New%20Initiatives.pdf

————. 2011. *From Representation to Inclusion: Diversity Leadership for the 21st-Century Military.* Arlington, VA.

National Academies (National Academy of Sciences, National Academy of Engineering, and Institute of Medicine). 2007. *Rising Above the Gathering Storm: Energizing and Employing America for a Brighter Economic Future*. Washington, DC: The National Academies Press.

——— (National Academy of Sciences, National Academy of Engineering, and Institute of Medicine). 2011. *Expanding Underrepresented Minority Participation: America's Science and Technology Talent at the Crossroads*. Washington, DC: The National Academies Press.

——— (National Academy of Engineering and National Research Council). 2012. *Assuring the U.S. Department of Defense a Strong Science, Technology, Engineering, and Mathematics (STEM) Workforce*. Washington, DC: The National Academies Press.

Office of National Drug Control Policy. 2010. *Progress Report on the National Youth Anti-Drug Media Campaign for Fiscal Year 2010*. Washington, DC. As of June 10, 2013:
http://www.whitehouse.gov/sites/default/files/ondcp/prevention/campaign_ effectiveness_and_rigor_page_youth_anti-drug_media_campaign__progress_ report.pdf

Office of the Under Secretary of Defense (Comptroller)/Chief Financial Officer. 2012. *Operation and Maintenance Overview, Fiscal Year 2013 Budget Estimates*. As of December 10, 2012:
http://comptroller.defense.gov/defbudget/fy2013/fy2013_OM_Overview.pdf

Oken, C., and B. Asch. 1997. *Encouraging Recruiter Achievement: A Recent History of Military Recruiter Incentive Programs*. Santa Monica, CA: RAND Corporation, MR-845-OSD/A. As of June 24, 2013:
http://www.rand.org/pubs/monograph_reports/MR485.html

Public Law 110-69. 2007, August 9. America Creating Opportunities to Meaningfully Promote Excellence in Technology, Education, and Science Act of 2007 (America COMPETES Act).

Public Law 111-358. 2011, January 4. America COMPETES Reauthorization Act of 2010.

RAND National Defense Research Institute, B. D. Rostker, S. D. Hosek, J. D. Winkler, B. J. Asch, S. M. Asch, C. Baxter, N. Bensahel, S. H. Berry, R. A. Brown, L. Werber, R. L. Collins, C. R. Cook, A. B. Cross, R. E. Darilek, N. K. Eberhart, J. G. Goulka, C. Gventer, A. Haddad, P. Heaton, W. M. Hix, E. V. Larson, R. J. MacCoun, S. O. Meadows, M. Pollard, E. Ratner, G. Ridgeway, J. Saunders, T. L. Schell, A. G. Schaefer, E. Wilke, and S. Young. 2010. *Sexual Orientation and U.S. Military Personnel Policy: An Update of RAND's 1993 Study*. Santa Monica, CA: RAND Corporation, MG-1056-OSD. As of June 10, 2013:
http://www.rand.org/pubs/monographs/MG1056

Riche, M. F., A. Kraus, and A. K. Hodari. 2007. *The Air Force Diversity Climate: The Air Force Diversity Climate: Implications for Successful Total Force Integration.* CNA Corporation.

Ruggles, S., J. T. Alexander, K. Genadek, R. Goeken, M. B. Schroeder, and M. Sobek. 2010. 2010 American Community Survey data. Edited by University of Minnesota, Integrated Public Use Microdata Series, Minneapolis, MN.

U.S. Census Bureau. 2008. 2008 National Population Projections. Website. As of June 10, 2013:
http://www.census.gov/population/projections/data/national/2008.html

U.S. Census Bureau, Bureau of Labor Statistics. 2011. 2010 Census Occupational Classification: Major Occupational Groups and Detailed Occupations Used in the Current Population Survey Beginning January 2011. As of June 11, 2013:
http://www.bls.gov/cps/cenocc.pdf

U.S. Department of Defense. 2012. *DoD Diversity and Inclusion Strategic Plan 2012–2017.* Washington, DC. As of June 10, 2013:
http://diversity.defense.gov/docs/DoD_Diversity_Strategic_Plan_%20final_as%20 of%2019%20Apr%2012[1].pdf

U.S. Department of Defense, Research and Engineering Enterprise. 2012. Mission history. Web page. As of June 10, 2013:
http://www.acq.osd.mil/chieftechnologist/mission/history.html

U.S. Department of Defense, Science, Technology, Engineering, and Mathematics (STEM) Executive Board. 2012. *STEM Strategic Plan, FY 2013–FY 2017.* Washington, DC.

U.S. Office of Personnel Management. 2011. *Government-Wide Diversity and Inclusion Strategic Plan 2011.* Washington, DC. As of June 10, 2013:
http://www.opm.gov/policy-data-oversight/diversity-and-inclusion/reports/ governmentwidedistrategicplan.pdf